SEEPAGE and LEAKAGE from DAMS and IMPOUNDMENTS

Proceedings of a symposium sponsored by the
Geotechnical Engineering Division
in conjunction with the
ASCE National Convention,
Denver, Colorado

May 5, 1985

Edited by Richard L. Volpe and William E. Kelly

Published by the
American Society of Civil Engineers
345 East 47th Street
New York, New York 10017-2398

The Society is not responsible for any statements
made or opinions expressed in its publications.

PREFACE

All major dams and impoundments are usually analyzed to determine the magnitude and location of seepage that can occur under, around or through the structure. More often than not, seepage protection elements within the dam or impoundment, in the form of geotextiles, natural filters and drains, are included in the design in order to control, collect and safely discharge the collected fluids. Darcy's Law has been successfully used by civil engineers for many decades for problems involving saturated flow, especially for the design of water retention dams. More recently, however, a new class of problems that deal with partially saturated leakage from waste impoundments has evolved. This latter class of problem can pose a serious environmental hazard if the leachate is toxic and it is not effectively collected and treated. The opportunity to discuss the current state of practice and to hear from several prominent engineers and researchers in the field of seepage and leakage from dams and impoundments occurred at the ASCE National Convention in Denver, Colorado during May, 1985.

Two committees of the Geotechnical Engineering Division of ASCE participated in sponsoring the symposium; namely, the Embankment Dams and Slopes and the Environmental Concerns committees. A Call for Papers went out in June 1984 and authors were invited to submit formal manuscripts for review and possible inclusion in this special volume. All papers were reviewed in accordance with the same procedures used by the Publications Committee of the Geotechnical Engineering Division. Members of the two committees, as well as at-large ASCE members, reviewed the papers. On the behalf of the entire ASCE membership, the editors wish to thank all reviewers for their efforts.

The session on Seepage From Embankment Dams was organized by Mr. Richard L. Volpe in his role as Chairman of the Embankment Dams and Slopes Committee. The collection of papers submitted on this topic provide insight to contemporary design considerations and approaches, as well as post-construction observation and evaluation techniques, and should be considered an excellent reference for both practicing engineer and student.

The session on Leakage From Impoundments was organized by Professor William E. Kelly in his role as Chairman of the Environmental Concerns Committee. The papers submitted on this topic, especially those dealing with partially saturated flow problems and the potential for contaminant migration, are considered major new contributions to the engineering literature.

Discussion of any paper contained in these proceedings is both invited and encouraged. Those readers wishing to do so should submit a formal discussion to ASCE for publication in the Journal of the Geotechnical Engineering Division. The format for such discussion can be found in any volume of the Journal. The closing date for submission is December 1, 1985.

Richard L. Volpe,
Geotechnical Consultant
Los Gatos, CA 95030

William E. Kelly,
University of Nebraska
Lincoln, NE 68588

Editors

CONTENTS

SEEPAGE FROM EMBANKMENT DAMS

SEEPAGE FROM IMPOUNDMENTS

Filters and Leakage Control in Embankment Dams

James L. Sherard, F. ASCE*
Lorn P. Dunnigan, M. ASCE**

Abstract

In the practice of earth dam engineering there has been some doubt that downstream filters could be relied upon completely to control and seal concentrated leaks through impervious dam sections (cores). This doubt led to the "multiple lines of defense" concept, particularly for high, central core dams, in which several independent lines of defense are provided to prevent concentrated leaks through differential settlement cracks and to control the leaks, if they should develop in spite of efforts to prevent them. This situation is currently changing. Recent laboratory research shows conclusively that filters with properties similar to those used in current common practice will reliably control and seal concentrated leaks. This conclusion is supported by experience with dam behavior. This increased confidence in filters is stimulating changes in practice, especially with regard to design and construction of the impervious earth core, frequently allowing lower cost. Recommended design criteria are given for filters needed to control concentrated leaks, based on results of a recently completed, 4-year laboratory research program carried out by the Soil Conservation Service.

Introduction

This paper presents a description of changing practice and opinions regarding design measures in embankment dams for leakage control. For such important structures as dams, changes in practice are only accepted gradually and remain debatable for a number of years. The main opinions presented regarding changes in practice are currently controversial but they have been generally accepted and applied by the engineering teams responsible for a considerable number of major dams in recent years. Hence, while the points presented may seem moderately surprising to some readers they have already been tempered by a number of years of exposure and debate in the profession, and are widely accepted by many specialists. The paper considers primarily the problem of leakage through the dam itself, assuming an impervious and erosion resistant foundation.

---

*Consulting Engineer, San Diego, California
**Head, Soil Mechanics Laboratory, National Technical Center, Soil Conservation Service, U.S.D.A., Lincoln, Nebraska.

Several causes of concentrated leaks through embankment dams can be imagined, such as through inadequately compacted layers. However, for practical purposes the only likely cause of a concentrated leak through the impervious earth section of a reasonably well constructed dam is a crack. In low dams in arid climates, cracks from drying and shrinkage have been troublesome. but cracks from differential settlement are the primary cause of concentrated leaks which must be considered by dam designers. Differential settlement causes open cracks on the surface and also creates deeper within the embankment conditions of low stress which allow the water to open and enter cracks which were not in existence before the water in the reservoir rose above the level of the crack, an action generally called "hydraulic fracturing".

In a companion paper (11) prepared for this conference, current available experience concerning hydraulic fracturing is summarized. This experience leads to the conclusion that hydraulic fracturing is much more common than was previously considered likely. Concentrated leaks occur in most embankment dams of all types and sizes without being observed. This is somewhat surprising; however. the evidence available for this conclusion is now so strong that it must be accepted. There can only be some remaining questions about details.

Leaks caused by hydraulic fracturing usually do not cause erosion, either because the velocity is too low or because they discharge into effective filters. Occasionally these concentrated leaks have caused erosion damage in central core dams in which the filters provided were too coarse. and have led to breaching failures of homogeneous dams that had no filters.

Both experience with dam behavior and recent laboratory research show conclusively that concentrated leaks are reliably controlled and sealed by adequate filters, even under the most extreme conditions. In the past, designers were not wholly confident about the efficacy of filters designed according to current criteria and they were reluctant to rely on them completely as the sole line of defense against possible concentrated leaks.

This paper scrutinizes the changes in design practice for embankment dam leakage control, which are currently being generated by the gradual acceptance of the two main conclusions described above; i.e., (1) concentrated leaks develop through most embankment dams, even without large differential settlement, in spite of design efforts to avoid them, and (2) filters with appropriate properties can be relied upon with complete confidence to control and seal concentrated leaks.

The changes consist of less emphasis on the necessity for keeping the dam core watertight by avoiding cracks and more emphasis on the importance and details of the downstream filter. In the past practice the designer held as an axiom that:

> The impervious core is the most important element in
> the dam. As long as the impervious core remains in-
> tact, with no cracks or other concentrated leaks, the
> dam will be safe. Therefore, the primary and most
> important objective of the design is to provide measures

which will minimize the likelihood of a concentrated leak to the greatest extent possible.

Based on the current available experience the designer is now inclined
to see the situation differently:

> We have been deluded in the past thinking that the
> impervious sections of our dams remain intact.
> Evidence now shows that concentrated leaks commonly
> develop in well designed and constructed dams. It
> is now clear that the most important element in the
> dam is the filter (or transition zone) downstream
> of the core. By providing a conservative downstream
> filter, we can quit worrying about possible concen-
> trated leaks through the core.

Brief Review of Development of Current Practice

History of Concern about Cracking

Starting about 1950 the profession began to be actively concerned
that open cracks were seen to be developing in embankment dams, even
in well constructed dams without unusually large settlement (9).
Previous to this time it had been generally assumed that impervious
sections of embankment dams compacted near Standard Proctor Optimum
water content were sufficiently flexible and deformable that open
cracks would not occur.

Particularly in the period 1960-75 a lot of attention was paid to
potential leaks in embankment dam cracks, both by researchers and
practicing engineers. In a 1963 reference book on embankment dams, it
was found that only six articles in the literature contained most of
the information on cracking published at that time (12). From 1965 to
1975 many dozens of articles were published on various aspects of dam
cracking, generally expressing concern that the problem was not
adequately understood.

In 1967 the ASCE Committee on Embankment Dams published a list of
the ten outstanding problems concerning design and construction of embankment dams for which research was considered urgently needed (1).
The problem of cracking was rated second highest in priority on this
list. Similar concern developed internationally about the same time.
At the 1970 ICOLD meeting in Montreal one of the General Reports (7)
and a considerable number of other papers and discussions were devoted
to cracking.

"Multiple Lines of Defense"

This concern led to the adoption of various design measures for
the two basic purposes of (1) reducing the likelihood of differential
settlement cracking, and (2) sealing and controlling concentrated leaks
if they developed in spite of the design measures provided to prevent
them. The situation existing in the early 1970's as seen by many
specialist engineers was described by A. Casagrande as follows (3);

"... I have been given credit for having invented cracks

in dams ...; i.e., for imagining something that does not
happen in reality. But as the height of dams increased,
the frequency of cracking increased also, and cracking
is no longer a figment of my imagination. ... To ensure
safety, the designer will provide several lines of defense
against cracking. The most important defense lies in the
selection of materials which are self-healing, particularly
in a wide zone downstream of the core, and if at all
possible, also in the core. In addition, a designer may
employ such defensive measures as upstream arching of the
dam, widening of the core ... etc."

Since about 1970-75 it has been fairly widely accepted that for
important dams "multiple lines of defense" should be provided against
leaks in cracks. This idea was applicable especially for high dams in
steep-walled canyons or other situations where large differential
settlement was inevitable. The various design measures making up these
lines of defense fall into two main categories:

(1)   Methods to Reduce Differential Settlement and the
       Likelihood of a Concentrated Leak

       (a)   Excavation of rock abutments to achieve flatter slopes
              and reduce abrupt changes in slope.

       (b)   For central core dams, using upstream sloping impervious
              core instead of vertical core.

       (c)   Compacting the bottom half of the impervious earth core
              at relatively lower water content and to as high density
              as practicable, to reduce its compressibility.

       (d)   For the upper part of the impervious core, compacting
              at higher water content, to make it flexible and capable
              of following imposed strains without cracking.

       (e)   Using plastic clay instead of silty sand for impervious
              core, also because of flexibility.

       (f)   Using strip of impervious core at especially high water
              content at the abutment contacts, sometimes using also
              a more plastic clay for this strip. (This is used to
              1) obtain a good seal with the soft, relatively erosion
              resistant soil against the rock, and 2) allow the em-
              bankment to settle with respect to the rock abutment.)

       (g)   Arching embankment dams in plan, to increase the
              longitudinal stress when water pressure pushes the dam
              crest downstream.

(2)   Measures to Control Erosion by Concentrated Leaks in Cracks
       That Develop in Spite of Preventive Measures

       (a)   Use of conservative downstream filters (transitions)

to prevent leaks from carrying eroded soil out of the core.

(b) Use of plastic clay for impervious core, believed to be more resistant to erosion of concentrated leaks.

(c) Use of sand in zones upstream of the core, the sand would be washed into and fill concentrated leakage channels.

Multiple lines of defense were considered necessary because of lack of confidence that any one of the lines of defense was completely reliable. This conclusion that multiple lines of defense were required has been widely accepted and applied, but has always been less than completely satisfying for dam designers. Its use leaves hanging the unspoken question, "If we are not wholly confident of any single line of defense, how can we be completely confident of any combination?"

Doubt About Filter Reliability

The adoption of the multiple line of defense practice was stimulated at least partially by doubt that downstream filters could be relied upon with complete confidence to prevent erosion of concentrated leaks through cracks in the core. This doubt reflected two main groups of questions: (1) questions about the validity of the early laboratory research on which current filter design criteria are based; and (2) questions as to whether experience with dam performance supported the conclusion that filters can be relied upon.

In the early research that led finally to our current accepted filter criteria the laboratory tests generally were made by the "conventional filter test procedure" in which water is caused to seep gradually through the pores of the base (protected) soil specimen and discharge into the filter. Also, with a few exceptions the pioneer research programs (2, 5, 17) were carried out on filters needed for cohesionless, relatively pervious sands. This older research generally did not include many tests on impervious base soils because the results of conventional filter tests on impervious soils were inconclusive. When a clayey soil is tested in a conventional filter test using a reasonable hydraulic gradient (2 to 20), the quantity of water which seeps through the clayey soil base specimen and discharges into the filter is so small (drops of water per hour) that there is no tendency for erosion. The significance of such tests was considered problematical at best.

As part of the design effort on specific projects, many individual engineers have made laboratory tests in which concentrated leaks were caused to flow through the impervious soil and discharge into the filter. The results of these tests are not generally published and there is no useful summary available in the literature.

As discussed later, the performance of dams with reasonable downstream filters has been generally very good. This was widely considered favorable experience supporting the conclusion that filters are successful; however, it was also realized that this generally good

experience could be due primarily to the fact that no concentrated
leaks developed in the dams; i.e., perhaps the filters in our dams are
not tested by concentrated leaks?

SCS Filter Research

The Soil Conservation Service is just completing a 4-year comprehensive research program devoted primarily to understanding the action
of downstream filters for controlling concentrated leaks in dam cores.
The work was carried out by SCS's National Soil Mechanics Laboratory,
Lincoln, Nebraska.

In this paper, we are discussing only the role of the filter in
sealing leaks and preventing erosion of the protected soil; i.e., we
are not discussing the needed permeability of the single band filter
which also functions as a drain.

In this research program a large effort was devoted to experimen-
tation with different types of laboratory tests having concentrated
leaks. Several different tests were employed using the same soil types
allowing an evaluation of the influence of the test procedure.

SCS researchers concluded that downstream filters with properties
not greatly different from those used in current common practice would
reliably control and seal concentrated leaks, even when the severity
of leakage was very much more severe (higher gradients and velocities
of flow) than could possibly exist in a dam. It was also concluded
that one of the several different types of tests employed was superior
for evaluating critical downstream filters.

Results of the first 2 years of the SCS research effort, devoted
primarily to fine silts and clays, were published in 1984 (13). During
the second 2 years the effort was broadened to include tests on filters
for coarse-grained impervious soils, but also substantial advances were
made in the understanding of filters for fine clays and silts.

Technical papers summarizing the results of the entire 4-year long
program are in preparation. Because these summary papers will probably
not be published for at least another 2 years, we are presenting here
the main results in brief form without supporting data. We expect no
fundamental change in the finally published results.

The paper describing the first 2 years of research on filters for
silts and clays presented two different types of laboratory tests --
slot tests and slurry tests (13). In both of these tests there was a
small amount of erosion and penetration of fines into the successful
filter before the filter sealed. Later in the research program another
test was adopted in which it was found possible to define a filter
boundary size (separating successful and unsuccessful tests) at which
"no visible erosion" of the walls of the leakage channel through the
base specimen took place during the test. Tests with filters slightly
coarser than the boundary had visible erosion. This test defined a
somewhat finer filter boundary than was defined for the same soils by
the previous slot and slurry tests. For example, for a common fine

clay in which the slot and slurry tests defined a filter boundary of
1.5mm (13), the new test gave a filter boundary of 0.5mm. This test,
which we have named the No Erosion Filter Test, was also found to work
very well for coarse-grained impervious soils, whereas the slot and
slurry tests were not satisfactory for impervious soils with d_{85} exceeding about 0.1 mm.

No Erosion Filter Test Procedure

The details of the laboratory test set-up are shown in Fig. 1.
The test procedure and criteria for judging the results are as follows:

(1) After preparation of the sample as shown in Fig. 1, with
cylinder in the vertical position for downward flow, fill the voids in
the gravel above the base specimen with water, letting the air escape
through the open hole in the top plate where pressure gage attaches.

(2) Attach pressure gage and open the valve to water source,
applying full pressure (about 4 kg/cm^2) to top of specimen rapidly.
For most soils water from the municipal supply was used. For dispersive clays, distilled water pushed by compressed air was used.

(3) Observe behavior for 5 to 10 minutes; this includes measuring
of quantity of water coming through the filter and observing of turbidity (color of water).

(4) Shut off water flow, dismantle apparatus, and observe erosion
of base specimen.

For each impervious (base) soil a series of these tests are made
using filters (sands and gravelly sands) with gradually increasing
coarseness, ranging from filters fine enough to prevent all visible
erosion of the hole in the base specimen through coarse filters that
permit considerable erosion. As previously reported (13, 14), it was
a primary and definite conclusion of the SCS research program that the
D_{15} size of the filter is a very good quantitative measure of the pore
size of the compacted filter that allows particles to pass through and
of its ability to act as a filter for base soils of different coarseness. Hence, normal (reasonably well graded) sand and gravel filters
are well defined by the single parameter (D_{15}), in millimeters. The
D_{50} size is not a good measure of the pore diameters of filter ability
of a sand or gravelly sand and filter criteria using D_{50} can be
abandoned.

These tests always allow definition of a "boundary" filter (D_{15b})
for any given base soil. For all tests on a given base soil with
filters finer than D_{15b} there is no visible increase in the diameter
of the initial hole through the base specimen. There is a very slight
erosion of base soil that seals the face of the filter quickly but the
quantity is too small to be observed. For tests with filters progressively coarser than the boundary (D_{15b}) the initial preformed
leakage channel (hole) in the base specimen, Fig.. 1, is eroded
progressively larger; however, for tests on fine-grained soils using
filters only slightly coarser than the boundary, with D_{15} in the range

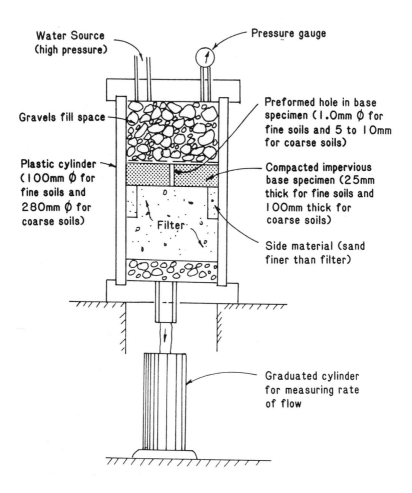

Water Source
(high pressure)

Pressure gauge

Gravels fill space

Preformed hole in base
specimen (1.0mm ∅ for
fine soils and 5 to 10mm
for coarse soils)

Plastic cylinder
(100mm ∅ for
fine soils and
280mm ∅ for
coarse soils)

Compacted impervious
base specimen (25mm
thick for fine soils and
100mm thick for
coarse soils)

Filter

Side material (sand
finer than filter)

Graduated cylinder
for measuring rate
of flow

Fig. 1:   No Erosion Filter Test Details
          (No Scale -- Schematic)

between 2 D_{15b} and 4 D_{15b}, the filter is generally sealed rapidly and completely during the test without a large amount of erosion (a few grams of base material seal the filter face).

For tests on sandy or gravelly sands with less than about 20% fines (passing the No. 200 sieve), the boundary D_{15b} was determined to be abrupt. In tests with filters only slightly coarser than D_{15b} a large amount of erosion occurs.

After much consideration of the results of tests of this kind performed on impervious soils of widely different properties, the following main conclusions were reached:

(1) The tests give reliably reproducible results and define a relatively narrow boundary between successful and unsuccessful filters;

(2) For most fine-grained soils the filter boundary defined (D_{15b}) gives a conservative but still reasonable filter size for direct use in design. It has a built-in safety factor because filters of this size essentially prevent all visible erosion of the specimen under the most severe conditions of laboratory testing, and because filters somewhat coarser also are sealed in the tests with only small erosion of the base specimen. The filter boundary (D_{15b}) defined by this test for typical silts and clays is roughly 20 to 40% of the boundary (D_{15B}) defined by the previously used slot and slurry tests (13).

(3) For coarse impervious soils, with less than about 20% of fines passing the No. 200 sieve, the boundary (D_{15b}) defined by the laboratory tests is so abrupt that it is desirable to apply a safety factor for use in design.

(4) The test results are influenced only by the grain size distribution of the base soils tested. The results are not influenced by the inherent erosion resistance of the base soil, as related to plasticity of the fines or potential for dispersive erosion, because the erosion conditions imposed in the laboratory tests (velocity, pressure and gradient) are so severe that all soils erode.

Recommended Design Criteria for Critical Filters

On any given job where it is desired to carry out laboratory tests to confirm the suitability of a proposed downstream critical filter, the test procedures and the evaluation criteria described above are considered suitable and appropriate. Testing using these procedures was carried out on a sufficient number of impervious soils, covering a wide range of coarseness and plasticity, to establish general criteria for evaluating critical filters for all types of soils used for the impervious sections of embankment dams.

The criteria depend only on the grain size of the impervious soil and are independent of the plasticity and of the dispersion potential of the soil fines. For gravelly impervious soils the gravel particles should be ignored in the application of the filter design criteria presented below (except as discussed below for gravelly sands of

Group 3). For soils with gravels a new particle size distribution
curve should be calculated for the portion of the soil that is finer
than the No. 4 sieve and this reconstructed gradation curve should be
used to apply the filter criteria. The criteria recommend below already include a conservative safety factor.

(1) **Impervious Soil Group 1 (Fine Silts and Clays):** For fine
silts and clays that have more than 85% by weight of particles finer
than the No. 200 sieve, the allowable filter for design should have
$D_{15} \leqq 9 d_{85}$ (where d_{85} is the size of the silt or clay for which 85%
is finer).

(2) **Impervious Soil Group 2 (Sandy Silts and Clays and Silty
and Clayey Sands):** For sandy (and gravelly) impervious soils with 40
to 85% by weight (of the portion finer than the No. 4 sieve) finer than
the No. 200 sieve, the allowable filter for design should have $D_{15} \leqq$
0.7mm. For these impervious soils, the influence of the fines
dominates in the filter test and the results are not influenced by the
existence of the sand particles.

The recommended criteria for this group of soils is different from
that presented in our earlier paper (13). Research carried out after
completion of the earlier manuscript showed that the difference between results of the No Erosion Filter Test and the previously used
slot and slurry tests was increasingly great for sandy impervious soils
with increasing d_{85} size. Using the slot and slurry tests the criterion $D_{15}/d_{85} \leqq 5$ appeared to be conservative, for soils in range of d_{85}
from 0.1 to 0.5 mm (13). Using the No Erosion Filter Test, it has now
become apparent that for soils in this range of gradation it is desirable to use a filter with D_{15} not exceeding about 0.7 mm as the
sample erosion for these soils becomes large before sealing with
coarser filters.

The recommended D_{15} size criteria for Groups 1 and 2 (D_{15} always
$\leqq$ 0.7 mm) apply to filters composed wholly of sand or gravelly sand in
which the sand fraction predominates and there is not a sufficient content of coarse gravel to cause excessive segregation during construction.
As discussed in more detail elsewhere (13), it is considered reasonable
and appropriate to require that the coarsest allowable gravelly sand
filters for fine-grained soils should have more than 40% sand sizes
(smaller than the No. 4 sieve) and maximum gravel size not exceeding
about 50mm.

(3) **Impervious Soil Group 3 (Sands and Sandy Gravels with Small
Content of Fines):** For silty and clayey sands and gravels with 15% or
less by weight (of the portion finer than the No. 4 sieve) finer than
the No. 200 sieve the allowable filter for design should have
$D_{15} \leqq 4 d_{85}$, where d_{85} can be the 85% size of the entire material including gravels. For soils in this category we have tested filters
only up to about $D_{15b} = 30mm$, but we believe that the rule is applicable
for coarser base-filter combinations.

(4) **Impervious Soil Group 4:** For coarse impervious soils intermediate between Groups 2 and 3 above, with 15 to 40% passing the No.

200 sieve, the allowable filter for design is intermediate, inversely related linearly with the fine content and can be computed by straight line interpolation. As an example, for an impervious sandy soil with 30% of silty or clayey fines and d_{85} = 2mm, the allowable filter for design is in between the value of D_{15} = 0.7mm (for soils of Group 2) and D_{15} = 4 (2) = 8mm (for soils of Group 3), and is calculated:

$$D_{15} = \frac{40 - 30}{40 - 25} \ (8 - 0.7) + 0.7 = 3.6 \text{ mm} \qquad \text{(See also Fig. 2)}$$

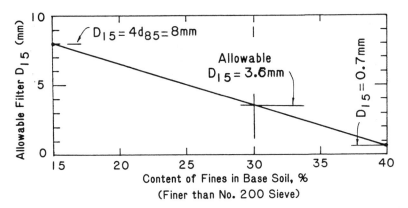

Example of Determination of Allowable Downstream Filter for Impervious Earth of Clayey Sand or Silty Sand with 30% Fines and d_{85} = 2.0 mm.

Fig. 2: Determination of Allowable Downstream Filter for Impervious Soil Group 4 (having between 15% and 40% finer than No. 200 Sieve).

Soil Groups 1 through 4 include all soils used for the impervious sections of embankment dams. The above criteria can be applied for all soils in Groups 1 and 2 regardless of the shape of the particle size distribution curve. For soils of Groups 3 and 4 the criteria apply to reasonably well-graded soils: for soils in Groups 3 and 4 which are highly gap-graded it is desirable to provide a filter for the finer portion of the gap-graded soil, or make No Erosion Filter Tests in the laboratory to select the appropriate filter.

Comparison with Current Criteria

Soil Group 1: For fine-grained silts and clays the criterion $\overline{D_{15} \leqslant 9 \ d_{85}}$ gives about the same filter in common use; that is, sand (or gravelly sand) filters with D_{15} of 0.2 to 0.7mm. Although there is no general rule widely quoted in the literature for filters for these fine soils, such sand filters are commonly used in practice.

Soil Group 2: For the sandy soils in Impervious Soil Group 2 above, the results of the laboratory tests were dominated by the fines, with filter boundary defined always of the order of 0.7mm, or a little larger. From the standpoint of behavior at the filter face, when tested as described above, these soils acted in the same manner as fine silts and clays which do not contain sand sizes (Soil Group 1). It was concluded from the observations of many tests on soils of this type that the fine particles dominate the behavior because the sand sized particles are simply floating in a matrix of the fines and it is the fine matrix which must be prevented from moving at the filter face in order to control safely and seal the concentrated leak.

For soils of Group 2 there is some question in the profession at the present time as to what filter criteria to use and how to apply the criteria. As an illustration, Fig. 3 shows the particle size distribution of a sandy clay, a soil representative of materials used for the impervious sections in many embankment dams all over the world. Soils of gradation similar to Fig. 3 are of all geologic origins from alluvium to weathered in-situ residual soils. In current practice there are two main criteria that would be widely considered for evaluating a filter for a sandy clay of this gradation:

Criterion 1 $D_{15}/d_{85} \leq 4$ (or 5)

Criterion 2 $D_{50}/d_{50} \leq 40$ (or 25)

Criterion 1 is the oldest criterion employed and is universally accepted in the profession. Over the years it was observed that Criterion 1 permitted filters that appeared visually to be coarser than considered reasonable for protecting some impervious soil types. This observation and some laboratory tests led to the proposal for Criterion 2 (16), which is currently accepted and applied widely but not universally.

As seen in Fig. 3, if only Criterion 1 were applied for this common sandy clay, it would allow a coarse gravel filter to be used, with $D_{15} \leq 4mm$. However, if Criterion 2 is applied it would require that $D_{50} \leq 0.24mm$; i.e., it would require a filter with more than 50% fine sand. It is seen, therefore, that when applied to evaluating a needed filter for this common sandy clay, Criteria 1 and 2 provide radically different answers. In common practice the filter allowed by Criterion 1 would be considered too coarse, but the $D_{50} \leq 0.24mm$ required by Criterion 2 would be considered unnecessarily fine and generally impractical.

In practice in this situation an intermediate filter is commonly used, such as a sand or gravelly sand with average D_{15} size not exceeding a maximum of the order of 0.5 or 1.0mm. Hence, the criterion based on the results of the recent SCS research, specifying a sand or gravelly sand with $D_{15} \leq 0.7mm$, is generally consistent with the current practice, and is intermediate between existing filter Criteria 1 and 2.

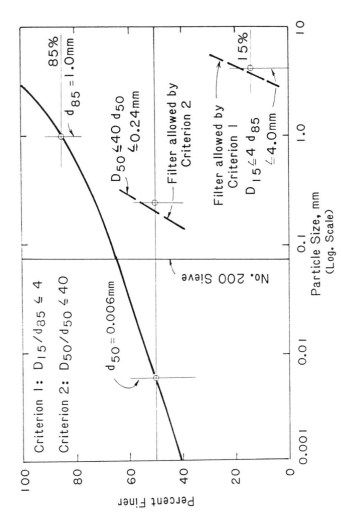

Fig. 3: Comparison of Filters for Sandy Clay as Required by Criteria 1 and 2.

In the practice of the last few years there has been a serious
question among specialists as to what filter is needed for broadly
graded impervious soils, such as those typical of glacial moraines,
which contain all particle sizes from cobbles down to clay sizes, with a
grain size curve that commonly plots as a near straight line on the conventional semi-logarithmic chart of particle size distribution. For
these soils there has been no general agreement on the appropriate way
to apply current filter criteria and there has been considerable trouble
with erosion of cores of these coarse soils in which coarse filters were
used (10). In the recent SCS research a considerable amount of effort
was spent testing these typical glacial moraines and similarly graded,
coarse, impervious soils. Results of the research show conclusively
that the particles coarser than fine sand in these soils cannot be relied upon to help seal the filter and that it is necessary to provide a
critical downstream filter that will retain the fines. The typical
coarse glacial moraines fall in Impervious Soil Group 2 and require a
sand or gravelly sand filter with $D_{15} \leq 0.7mm$.

Soil Group 3: For sandy soils with a small fine content, the criterion $D_{15} \leq 4 d_{85}$ is the same as the current practice. For these soils
the laboratory tests show the failure boundary generally at about
$D_{15b} = 7.5$ to 9, so that the design criteria ($D_{15} \leq 4 d_{85}$) has a safety
factor of the order of two.

All these filter tests and criteria are for "critical" filters,
which are located downstream of the dam core and are needed to control
and seal concentrated leaks. For upstream filters and other filters
placed adjacent to impervious embankment sections where it is not
possible for a concentrated leak to develop, coarser materials can be
safely used, as discussed later.

Experience with Filters in Dams Is Generally Good

In the last 30 years or so thousands of dams have been designed and
built with criteria and procedures not greatly different from the current practice. A small fraction have had behavior problems of various
kinds. Some of these problems never become known widely in the profession. Even for those cases which are thoroughly studied, with results presented in the literature, it is frequently difficult for the
investigators to be completely confident that the cause of trouble was
reliably understood.

In spite of these difficulties it is possible to extract from the
mass of experience with dam behavior some conclusions on which there
should be wide agreement in the profession. One of these is that there
are practically no records of leakage or internal erosion (piping)
through earth cores that is attributed to filters considered reasonable
according to current practice. Almost all records of damage or failure from leakage and piping have been for dams in two specific groups:
(1) homogeneous dams without chimney drains (vertical filter-drains), in
which a concentrated leak develops through the dam and exits on the downstream slope without passing through a filter; and (2) dams with impervious sections consisting of coarse, broadly graded soil for which ex-
cessively coarse filters were used (10).

As discussed above, there has been a major problem in evaluating
the "lessons learned" from experience with dam behavior. Problems with
erosion (piping) practically never have developed in dams provided with
reasonable downstream filters. However, it was not known whether this
good experience could generally be attributed to the action of the
filter in sealing concentrated leaks, or whether no leaks developed and
the filters were not tested by concentrated leaks.

In the last few years much evidence has been accumulating to support the conclusion that concentrated leaks commonly develop without
being observed, due to differential settlement. The evidence and experience is now so great that there is no longer any reasonable doubt:
concentrated leaks by hydraulic fracturing from differential settlement occur in most well designed and constructed dams, even those without unusual settlement. These leaks are not observed and usually their
existence is not suspected because they do not erode and they become
sealed before much water passes through, either because the velocity is
not high enough or because the leak discharges into an effective filter.
These unobserved leaks develop in cracks during or soon after the first
reservoir filling and then later the cracks close by swelling or softening and squeezing shut. The evidence that concentrated leaks by hydraulic fracturing are common, too extensive to present adequately
here, is described in another paper prepared for this Symposium (11).

In summary, concentrated leaks develop commonly. In dams of
moderate height or greater these leaks are sufficiently large to cause
erosion but they do not erode because the downstream filters (or
transitions) commonly provided in current practice are effective. The
writers have no doubt that the overall lesson learned from the mass of
experience available with dam behavior is that concentrated leaks commonly develop in well designed and constructed dams and that "filters
work satisfactorily" to control the concentrated leaks.

Impervious Earth Core Properties

In the past many engineers considered that clay was the superior
material for the impervious core. In some geographic areas this opinion was so strongly entrenched that the terms "core" and "clay core"
were practically synonymous. There are many examples in the last 20
years where clay has been hauled great distances (30 to 40 km or more),
at considerable cost although large uniform deposits of excellent cohesionless silty sand were available immediately at the dam site. Clay
was considered superior to cohesionless impervious materials because it
would be more deformable and better able to follow imposed strains from
differential settlement without cracking, and because it would have
higher resistance to erosion if a concentrated leak developed from any
source. Highly plastic ("tough") clay was particularly prized.

In the current trend, supported by increasing confidence in downstream filters, this preference for "clay" over cohesionless impervious
soils is not considered valid. The two materials act differently in the
presence of a concentrated leak in a crack. Each has theoretical advantages and disadvantages, and it is not clear that either is superior.
The tough, highly plastic clay will definitely resist erosion of the

walls of a leak in an open crack better than a cohesionless silty soil.
But the higher unconfined compressive strength of the clay core provides more likelihood that the embankment material can arch around the
leakage channel and keep it open. In impervious embankments of cohesionless silty sand the walls of initially open, water-filled cracks
quickly become saturated, soften and tend to squeeze shut.

The greater resistance of clay to erosion of concentrated leaks is
believed not important. All clays will erode under severe conditions
and it is not considered reasonable or conservative to rely on the relatively feeble erosion resistance of a clayey soil to provide defense
against piping of a concentrated leak. The principal job of the core
is to be impervious. Prevention of piping is the job of the filter.

In the past, sometimes a superior core material was manufactured by
blending together two materials of radically different properties. For
example, at some dams cohesionless fine silts (or fine clays) were
mixed with sandy gravel to manufacture a core material with the imperv-
iousness of the silt or clay and the high shear strength, low compressibility and "self-healing" properties of the sandy gravel. This has been
done only rarely, but for some major dams. The current trend would be
away from this practice and towards the use of the fine silt or clay
for the core and the sandy gravel in other zones of the dam, even for
high dams.

In summary, the current trend is to use the least costly of the
available impervious soils for a central core, although it may be completely cohesionless and relatively erodible when exposed to concentrated water flows, and may be relatively compressible when compacted in the dam.

As historical background for this point it is desirable to record
that the previous strong preference for impervious cores of plastic
clay was not at all universal. In some parts of the world there are
no plastic clays available for dam construction. In these areas many
major dams have been built over the last 30 years with impervious cores
consisting of cohesionless silty sands, sandy silts, gravelly, sandy
silts, etc. In some areas these materials (usually cohesionless silty
sand) have been largely decomposed granites or similar rocks which are
frequently considered among the best impervious earth materials available for embankment dams. In other areas great numbers of dams have
been built with impervious cores of glacial moraines, often in the form
of cohesionless gravelly silty sand. Impervious dam sections from
weathered sandstone, often in the form of very fine, uniform silty
sand, are common. A list of dozens of major dams built with cores of
completely cohesionless silty soils in the last 30 years could easily
be compiled.

Because of the above experience it may seem unnecessary to emphasize this point so strongly. Presentation of this subject is
needed in the literature at the present time because there have been
so many debates on this subject in connection with major projects
only in the last few years. It is necessary to record that some
specialized engineers of long experience have the very strong opposite

opinion that it is worthwhile to spend a lot of extra money to obtain a clay core.

An exception to the general practice described above of using the least costly impervious material available arises for the strip of impervious core material directly in contact with hard rock foundations. There is always concern that erodible impervious earth core material can be carried into cracks in the foundation rock, where there is no filter to prevent the erosion. Cracks in the rock surface are sealed by consolidation grouting and surface treatment (slush grout, shotcrete, etc), but there is always a possibility that some cracks are not adequately sealed and, for high dams at least, there is also a possibility that sealed cracks may open again in the treated rock surface by the foundation deformations caused by the weight of the embankment. There have been many cases where erosion of core material into rock cracks has caused serious problems.

For these reasons it is not desirable to place very erodible fine-grained soils (fine, cohesionless silts or dispersive clay) directly on rock foundations. For dams on rock foundations where fine, highly erodible soils are used for the main impervious core, it is desirable to place a thin layer of a different impervious material directly against the foundation rock surface, the material having properties so that it will not be easily eroded into small open rock cracks.

Earth Core Width, Slope and Compaction

The main decisions relating to the earth core have generally been influenced by the designer's concern over the risk of differential settlement cracks. For rockfill dams with central cores of compressible earth, the core is often made to slope upstream, instead of vertically. The upstream sloping core is generally considered to be less likely than the vertical core to develop low internal stresses from arching of the top between the upstream and downstream shells.

The minimum acceptable core width of a central core dam at a given site is often a subject for debate. The greater the potential differential settlement, generally the less likely that a very thin core would be used.

Particularly for high dams in narrow valleys, some engineers insist on making the lower half of the dam less compressible by heavy compaction at relatively low water content. Sometimes also the upper part of the core is compacted at a relatively high water content in order to make it flexible and able to follow the strains resulting from the differential settlement without cracking.

The current design trend, based on increased confidence in the reliability of downstream filters is changing practice on these points. The geometry and width of the core are generally chosen on the basis of other considerations. Less emphasis is placed on reducing differential settlement and cracking by varying the construction water content: the trend, even for high dams, is to compact the entire core at the same water content with the same compaction effort.

The width of the earth core in a dam with rockfill or gravel shells is generally chosen arbitrarily on the basis of precedent, commonly in the range between 30 and 60% of the head. There are some well known examples of completely successful dams with thinner cores; for example, at the 80 m. high Nantahala Dam, completed in 1939, in Tennessee, with sloping earth core, rockfill shells and an excellent downstream filter of sand, the width of the core was about 11% of the head (12). In the 45 years since the construction of Nantahala Dam there have not been many dams built with cores thinner than about 25% of the head.

In the last few years thinner cores have been used at a few dams with good filters. The Svartevann Dam in Norway, completed in 1976, is a rockfill dam about 130 m. high, with upstream sloping core of cohesionless glacial moraine (6). The core width was made about 18% of the dam height, using a good downstream filter of gravelly sand with average D_{15} about 0.1 mm. Because the Svartevann Dam is the highest dam in Norway it was provided with more instruments than commonly used to allow detailed understanding of its behavior, which is shown to be satisfactory in every respect.

The main function of the core is to make the dam relatively watertight. Most compacted impervious soils have a coefficient of permeability of 10^{-6} cm/sec., or less. Hence, theoretically a core width of 2 or 3 meters would be adequate to reduce the total quantity of seepage to a small value, even for relatively high dams.

In the laboratory it is shown conclusively that a compacted silt or clay specimen with thickness of only a centimeter or two, when it has an appropriate filter under it, will permanently and securely act as an impervious barrier to hold back water with pressures only limited by the capacity of the laboratory equipment. The writers have made many such tests with compacted specimens of various types of impervious soils about 2 centimeters thick with appropriate sand and gravelly sand filters, using pressures of about 4 kg/cm^2, giving an hydraulic gradient of about 2000. Nothing happens in these tests, even when holes are punched through the specimen causing a concentrated leak as shown in Fig. 1.

These experiences show clearly that as long as the water seeping through the earth core material discharges into an adequate filter, there is no practical limit on the tolerable hydraulic gradient that can be safely imposed on the impervious earth core. Therefore, it is reasonable in the future to contemplate making earth cores in dams much thinner than the minimum widths used in current practice.

Junctures Between Embankment Dams and Concrete Structures

Before about 1960-65 it was standard practice to provide cutoff collars on concrete conduits passing through the base of embankment dams, over the length of embedment in the impervious core. These were simply walls protruding out into the surrounding dam embankment. Similar collars were commonly provided on spillway walls and other concrete structures where embankment dams connected to them. The purpose was to force the particles of water that were seeping from upstream to downstream along the earth-concrete interface to have a longer path and not a straight line.

Over the years many engineers concluded that these cutoff collars had the strong disadvantage that they made it impossible to compact the impervious earth core directly against the concrete structure in a reliable fashion with heavy rollers. A large amount of compaction by hand labor was needed, which was clearly inferior. Because of increasing concern over this problem, and at the same time increasing confidence in the efficacy and reliability of filters, over a period of years the number and size of these collars were gradually reduced, and special attention was paid to the downstream filters. This trend continued until the collars have been eliminated altogether.

A common current practice is generally as follows:

(1) No seepage collars provided.

(2) The exterior concrete surfaces at the concrete-earth core interface are made smooth and are sloped slightly off vertical (such as about 1H:8V or 1H:10V) so that the earth can be compacted directly against the concrete by the wheels of a rubber-tired vehicle running parallel with the concrete face, Fig. 4. The earth core is placed against the normal smooth, concrete surface; i.e., the concrete surface is not sandblasted, chipped, painted or treated in any way.

(3) Special attention is paid to the downstream filter. The filter is often wrapped completely around the downstream portion of conduits; i.e., underneath as well as on both sides and the top, so that all potential leakage water travelling along the concrete-earth core interface must exit in a controlled fashion in the filter.

The above practice has been used successfully at many projects. One interesting example is at the large Guri Hydroelectric Project, recently completed in Venezuela, where a 125-meter-high central core rockfill dam is connected to a concrete gravity dam, Fig. 5. The smooth concrete face of the gravity dam over the area of contact with the earth core was made at 7V to 1H, and the earth was compacted against the concrete as shown in Fig. 4. The angle between the longitudinal axis of the rockfill dam and the face of the concrete dam was made 80° (instead of 90°), with the idea that downstream deflection of the dam under the push from the reservoir water would wedge the core against the concrete. On similar connections on other jobs the earth core has been brought in at 90°, which the writers believe is also reasonable.

At the concrete face the downstream critical filter at the Guri connection was made several meters wider than the normal width in the rockfill dam, a common desirable detail. For embankment dams in which the critical downstream filter contains any significant quantity of gravel-sized particles, there is a danger that the gravels could segregate at the concrete-earth core interface forming pockets of gravel without sand in the voids. In this situation, in addition to increasing the width of the filter at the concrete surface, it is desirable to consider using a thin layer of filter material consisting only of sand at the interface.

Fig. 4: Compaction of Impervious Clay Core Against Smooth
         Sloping Concrete Surface with Wheels of Heavy,
         Rubber-Tired Vehicle.

Fig. 5: Connection Between Rockfill Dam and Concrete Dam
         (Guri Hydroelectric Project) Showing Smooth
         Concrete Surface Under Impervious Earth Core.

High Dams in Narrow Canyons

Starting about 1950 when increasing concern was developing about cracking, there was debate in the profession about the maximum permissable steepness of rock abutments. There was no general agreement on this point, but some engineers in those years held the opinion that rock abutments steeper than about 1:1 were probably undesirable. The steeper the abutments, the narrower the canyon and the higher the dam, the more concern was felt about the suitability of an embankment dam, and the more likely a concrete dam would be chosen.

This concern over high steep rock abutments has been gradually relaxing as good experience has accumulated and as confidence has developed in the reliability of filters. In the last 30 years, an increasing number of embankment dams have been built successfully with increasing height in canyons with increasingly steep rock valley walls. At the present time very high embankment dams are being built in narrow, steep-walled canyons where only concrete dams would have been considered acceptable previously.

Two notable examples of the current practice are the Chicoasen Dam in Mexico (8) and the Chivor (Esmeralda) Dam in Colombia (15). The Chicoasen Dam, a central core rockfill dam, 260 m. high, completed in 1980, has nearly vertical rock abutments over most of the dam height. The 240 meter high Chivor Dam, a sloping earth core rockfill dam completed in 1975, was built in a narrow, rockwalled canyon with width of only about 320 m. at top (crest length). The behavior of the Chivor Dam as recorded by an extensive system of instrumentation has been completely normal in all ways (4). At present in Colombia at a site not far from the Chivor Dam, the Guavio Dam, designed to have a maximum height of 250 m., and crest length of about 360 m., is being built in another steep-walled canyon.

This trend will probably continue until it is generally agreed that the highest embankment dams can be safely built with earth cores in narrow canyons with near-vertical abutments. It is only necessary that the rock canyon walls are not overhanging. This conclusion is based firmly on confidence in the reliability of filters to control and seal concentrated leaks.

Arching Dams in Plan

Concern over cracking in the 1950's caused some engineers to design embankment dams with the longitudinal axis curved. The purpose was to cause the length of the dam to decrease if any downstream deflection occurred, thereby tending to close cracks.

In the current trend, curved axes are not considered necessary. Straight axes are considered satisfactory for even the highest embankment dams in narrow canyons. The Chivor, Chicoasen and Guavio Dams described above, all have straight axes.

Scarification of Compacted Layers and Slickensides

In current practice there are strong differences of opinion about

the need for, and amount of, scarification between compacted layers to
achieve bond, especially for impervious sections compacted with smooth
rollers. Also, however, modern tamping rollers with short feet of
fairly large area (c ompared to the older type sheepsfoot rollers with
long feet and small area) commonly leave the final compacted layer
surface relatively smooth and hard. The practice for bonding between
layers on various jobs in recent years has ranged from no scarification
(usually when the impervious soil being compacted is fairly wet and the
two layers are shown to fuse together naturally) to the use of several
passes of a disc harrow which cuts nearly through the entire thickness
of the previously compacted layer.

Another common problem is the development of "slickensides" or
"laminations" in the compacted soil, caused by shearing movements under
the rubber tires of trucks or other heavily loaded equipment. This
occurs most frequently in fine clayey soils with water content near or
above Standard Proctor Optimum, but it can occur in heavily travelled
areas in embankments of all types of soils (clayey sands, etc.).
Troubles have been caused on quite a few jobs where these slickensides
develop and it is considered necessary to stop the work and study the
problem. The typical conclusion from these studies is that the
slickensides have no significant influence on thepermeability. The
shear strength of the mass may be decreased moderately but this is
usually not an important factor in a rockfill or gravel fill dam on a
sound rock foundation.

For the central core dam the worst conceivable result of not
scarifying to obtain bond between compacted layers, or of slickensides
within compacted layers, is a condition in which it can be imagined that
a leak by hydraulic fracturing could develop more easily than otherwise.
In the current trend in practice with increased confidence in the
reliability of downstream filters there is much less concern about inadequate bonding between layers and slickensides, and less effort and
money is spent to eliminate them.

The writers believe that in future practice routine scarification
of the top of compacted layers can be abandoned, even using smooth
steel-drum rollers. If the soil is reasonably moist the two layers will
bond together adequately without scarification. If the top of the
previous layer has dried out, satisfactory bonding can be obtained by
sprinkling the surface before the next layer is placed on top.

Upstream Sand Zones as "Crack Fillers"

A fairly common design detail in central core dams is a sand zone
(filter) 2 to 4 m. wide at the top of the dam on the upstream face of
the impervious core. The purpose is to provide a reservoir of sand
that would be washed into a crack that might develop through the upper
part of the impervious core from any cause, such as differential settlement or earthquake. This detail was developed as one of the multiple defense lines, and has been used fairly commonly since about 1965.

In order to serve the intended purpose, the maximum particle size
of the sand needs to be smaller than the width of the crack that might

develop. Hence, sand with a maximum size not larger than about 1.0 mm has commonly been considered desirable, so that it could reliably be washed into cracks with widths of several millimeters without bridging over and blocking the entrance. In some cases designers have provided coarser sand, such as with maximum size of 6 mm., or even 12 mm., considering that it might be useful if a major crack opened.

The potential value of this upstream sand "crack filler" is debated at the present time, with strong opinions on both sides. In recent years many important central core dams have been built without these zones, even though suitable sand was available at reasonable cost.

The several main arguments raised against this upstream sand zone are:

(1) The filter downstream of the core is the main line of defense for controlling and sealing concentrated leaks, and it must be relied upon. Hence, if there is any question about the adequacy of the downstream filter, it should be made more conservative and the question eliminated, rather than providing another line of defense of questionable efficacy upstream.

(2) It is difficult to be confident that the upstream sand would fill the leakage channel. The velocity of flow in a concentrated leak, even in a fairly wide open crack, is not likely to be very high because it is limited by the permeability of the downstream filter, the upstream face of which should be progressively clogged by eroded debris from the core.

(3) There is an argument as to whether it is desirable for the upstream sand to enter a crack, because a limited amount of sand may enter and tend to prop the crack open and tend to prevent it from squeezing shut by softening or swelling of the crack walls.

(4) In addition to the unit cost of the upstream sand zone itself, the sand zone upstream usually requires another band of coarser transition to hold the sand in place, which complicates the construction and adds more cost.

The writers believe that this upstream sand zone will be used less frequently in the future and probably will be finally abandoned as a design practice.

Upstream Transitions for Clay Cores

One of the most common dam types built in the last 30 years has been the central core rockfill dam, often with a clay core. In such a dam there are commonly two or three zones of progressively coarser filters (sometimes called transitions) between the core and the downstream shell. Twenty years ago the common design was to make the filters upstream of the core the same as those downstream; i.e., the same number of zones and gradation of materials and the same widths of zones.

In the last 20 years there has been a gradual trend toward making the upstream filters in these dams more economical. This was motivated by increasing realization that the upstream filters serve a much less important role than the downstream filters.

The main function of the upstream filter is simply to prevent the clay core material from migrating in the upstream direction into the voids in the upstream rockfill shell and, theoretically, to act as a filter for the core following reservoir drawdown. In fact, consolidated under the pressure inside the dam the clay core is in a relatively stiff state, so that there is no tendency for it to simply migrate, by falling under gravity action in the upstream direction. Also, for an impervious clay core, the amount of water that seeps out of the core in the upstream direction following reservoir drawdown is very low. The maximum gradient for the seepage in the upstream direction following drawdown is less than 1.0, and the velocity and energy of the seepage entering the upstream filter is too small to erode clay particles and carry them upstream. Such an upstream filter is considered a noncritical filter because it is never called upon to act to control a concentrated leak.

As a part of the recent SCS filter research tests were made with relatively erodible clays of low plasticity (using a gradient of about 20 and very coarse filters). These tests showed clearly that it is not necessary to apply the current filter criteria to such noncritical filters as the upstream filter on a central core dam (13). Hence, the laboratory research results support the trend in practice to make the upstream filter more economical.

The first steps in this trend resulted in making the upstream filters narrower than the downstream filters. The trend is continuing and the current designs for a considerable number of high dams retaining very large and important reservoirs have only one coarse transition between a very fine clay core and the upstream rock fill shell. This transition is commonly a zone of grizzlied or crushed rock, such as from 150 millimeters maximum size down. The relationship between the particle sizes of such a small rock transition zone and a fine clay core is completely outside our current criteria for acceptable filters.

As an example, Fig. 6 shows the cross-section of the 100-m high rockfill dam used in the main river section for the large Tucurui Hydroelectric Project. recently completed in Brazil. The earth core is an extremely fine clay of medium to high plasticity. As seen, the design has four bands of filters and transitions downstream and only one band of grizzlied rock upstream in the lower part of the dam. Fig. 7 shows the typical appearance during construction of the contact between the clay core and the upstream zone of grizzlied rock. Inevitably there is segregation and the upstream zone of small rock in no way can be considered a filter for the fine clay core: it is simply acting to retain the upstream face of the stiff clay core. The clay is prevented from migrating upstream into the voids of the grizzlied rock zone because it has sufficient stiffness to arch across the voids. The clay core is stiff already when it is compacted (unconfined compressive strength of the order of 3 kg/cm^2) and then it becomes stiffer when consolidated under the weight of the overlying embankment.

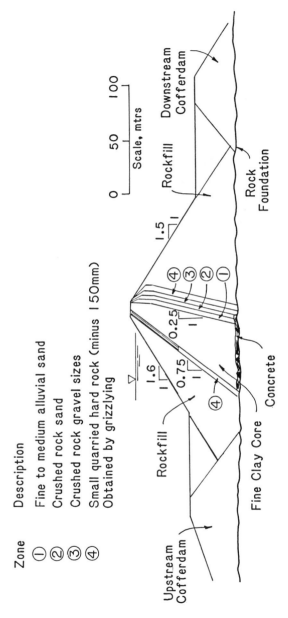

Fig. 6: Cross-Section of Tucurui Rockfill Dam in Main River Section Showing Difference Between Upstream and Downstream Filters

Fig. 7: Typical appearance contact between fine clay core and upstream
         transition of small (grizzlied) rockfill during construction
         at Tucurui Dam (see Fig. 6). Notebook in foreground is
         18 by 25 cm.

The above conclusions are valid for clayey cores because their unconfined compressive strength and low permeability eliminate any significant tendency for the material forming the upstream face of the
core from migrating in the upstream direction. We believe the conclusion applies to any core material that has 30% or more of fines
passing the No. 200 sieve) of clay or cohesive silt with plasticity
index of 8 or higher.

For cores of cohesionless silty sands and sandy silts, the material
of the upstream face of the core can be reasonably imagined to migrate
upstream into transitions with large voids, under the combined action
of gravity and seepage following reservoir drawdown. Especially for
dams retaining reservoirs that are frequently lowered (such as pumped
storage reservoirs in the extreme), and for dams with thin, near-
vertical central earth cores, with steep upstream faces, it is desirable
to use a reasonably conservative upstream filter.

Summary and Conclusions

    1. Concentrated leaks by hydraulic fracturing occur in most em-
bankment dams without being observed, even in dams without large
settlements. Usually the leaks do not cause erosion either because the
velocity is too low or they discharge into effective filters.

    2. Both recent extensive laboratory research and evaluation of ex-
perience with dam behavior support the conclusion that adequate filters
will reliably seal and control concentrated leaks through earth cores

of embankment dams.

3. A main conclusion from the recently completed, 4-year long SCS filter research program is that the No Erosion Filter Test, described herein, is superior to the slot and slurry tests used earlier in the research for silts and clays (13). It was found to be the best test for the general filter research and also the best available test for routine laboratory evaluation of filters for specific projects. It is applicable for tests on coarse impervious soils as well as fine clays and silts.

4. Filters needed for control of concentrated leaks through impervious dam cores are not greatly different than those used in current common practice. Design criteria for various impervious soils are shown to be primarily dependent on the content fines (passing the No. 200 sieve) in the following main categories:

Soil Group	% Fines* ($\angle$ No. 200 Sieve)	Recommended Design Criteria (already incl. adequate safety factor)
1	85 to 100%	$D_{15}/d_{85} \leqq 9$
2	40 to 85%	$D_{15} \leqq 0.7$ mm
3	0 to 15%	$D_{15}/d_{85} \leqq 4$
4	15 to 40%	Intermediate between Groups 2 and 3

*Where % Fines is the fine content of the sand fraction ($\angle$ #4 Sieve)

5. Increased confidence in filters is causing a change in practice. Filters are being relied upon to seal concentrated leaks and less money and effort is being spent on the various design measures formerly provided to reduce differential settlement.

6. Impervious soils with cohesionless silty fines are equally satisfactory for impervious core materials as clayey soils. As a general rule the least costly available impervious soil should be used for the dam core.

7. Central vertical cores are equally satisfactory as upstream sloping cores, even for high dams in steep canyons.

8. High embankment dams can be safely built in very narrow canyons with steep, near vertical, rock abutments.

9. Dams with straight longitudinal axes are as satisfactory as dams that are curved upstream in plan.

10. It is reasonable conservative practice to compact the entire impervious section of a high dam at the same water content for the full height. It is not necessary to compact the lower part at higher density and lower water content or the upper part at higher water content.

11. Seepage collars on conduits and at connections with concrete dams are no longer used.

12. Upstream sand zones provided for the purpose of being carried into and sealing cracks are not favored.

13. The increasing confidence in filters allows consideration of thinner earth cores in the future practice.

14. At sites where relatively high differential settlements are inevitable, the most effective method of ensuring a high margin of safety for the dam is to increase the conservatism of the critical downstream filter.

15. For typical coarse glacial moraines, graded from cobbles to fines, and other similarly graded impervious soils, there has been an unresolved question in the profession about the needed filter. The recent SCS research concentrated on this problem, concluding that a sand or gravelly sand with $D_{15} \leqq 0.7$ mm., is needed for a conservative downstream filter.

16. Adequate bonding between compacted layers can be obtained without the need for routine scarification of the top of the previously compacted layer.

17. For the transition between an impervious core of fine clay and an upstream rockfill shell, a single band of relatively coarse material is adequate, such as grizzlied hard quarried rock.

. . . .

Appendix I.--References

1. ASCE (1967), "Problems in Design and Construction of Earth and Rock-fill Dams," Journal of the Soil Mechanics and Foundation Engineering Division, ASCE, May, 1967, p. 129.

2. Bertram, G. E., "Experimental Investigation of Protective Filters," Soil Mechanics Series No. 7, Graduate School of Engineering, Harvard University, 1940.

3. Casagrande, A., "Transcript of Informal Discussion on Embankment Dam Design," 7th International Conference on Soil Mechanics and Foundation Engineering, Mexico, Vol. 3, 1969, p. 301.

4. Hacelas, J. E., and Ramirez, C. A., "Interaction Phenomena Observed in the Core and Downstream Shell of Esmeralda Dam, Chivor Hydro-electric Project," Specialty Conference Session 8, 9th International CSMFE, Tokyo, June 1977.

5. Hurley, H. W., and Newton, C. T., "An Investigation to Determine the Practical Application of Natural Bank Gravel as a Protective Filter for an Earth Embankment," Thesis, Dept. of Civil Engineering, Massachusetts Institute of Technology, 1940.

6. Kjaernsli, B., et al, "Design, Construction Control and Perform-
    ance of the Svartevann Earth-Rockfill Dam," 14th ICOLD, Rio de
    Janeiro, Vol. IV, 1982, pp. 319-337.

7. Lowe III, J., "Recent Development in the Design and Construction of
    Earth and Rockfill Dams," Proc. 19th ICOLD, General Report, Ques-
    tion 36, Montreal, Vol. V, 1970, pp. 1-28.

8. Moreno, E., and Alberro, J., "The Behavior of the Chicoasen Dam:
    Construction and First Filling," 14th ICOLD, Brazil, Vol. I, 1982,
    pp. 155-182.

9. Sherard, J. L., "Embankment Dam Cracking," Embankment Dam Engineer-
    ing, John Wiley and Sons, New York, 1973, pp. 272-353.

10. Sherard, J. L., "Sinkholes in Dams of Coarse Broadly Graded Soils,"
    13th ICOLD Congress, India, Vol. II, 1979, pp. 25-35.

11. Sherard, J. L., "Hydraulic Fracturing in Embankment Dams," Paper
    prepared for Symposium on Leakage Control in Embankment Dams, ASCE
    Denver, Colorado, May 1985.

12. Sherard, J. L., Woodward, R. J., Gizienski, S. F. and Clevenger,
    W. A., Earth and Earth-Rock Dams, John Wiley and Sons, New York,
    1963.

13. Sherard, J. L., Dunnigan, L. P. and Talbot, J. R., "Filters for
    Clays and Silts," Journal of the Geotechnical Engineering Division,
    ASCE, June 1984, pp. 701-718.

14. Sherard, J. L., Dunnigan, L. P. and Talbot, J. R., "Basic Proper-
    ties of Sand and Gravel Filters," Journal of the Geotechnical
    Engineering Division, ASCE, June 1984, pp. 684-700.

15. Sierra, J. M. and Buendia, J., "Esmeralda Dam, Chivor Project,"
    (Spanish) 5th Panamerican Conference on Soil Mechanics and
    Foundation Engineering, Buenos Aires, Vol. II, 1975, pp. 277-293.

16. U. S. Army Engineers, "Filter Experiments and Design Criteria,"
    Technical Memorandum No. 3-360, Waterways Experiment Station,
    Vicksburg, Mississippi, 1953.

. . . .

Appendix II--Notation

The following symbols were used in this paper:

d_{85} =   particle size in base soil for which 85% by weight of
              particles are smaller (similarly for d_{50}).

D_{15} =   particle size in filter for which 15% by weight of
              particles are smaller (similarly for D_{50})

D_{15b} = D_{15} size of filter found to be boundary between successful
              and unsuccessful tests for any given base soil, using

no-erosion filter tests.

D_{15B} =   Similar filter boundary for a given base soil as determined by the slot and slurry tests used earlier in the
research program (13).

EMBANKMENT SEEPAGE CONTROL
DESIGN AND CONSTRUCTION

By Edward C. Pritchett,[1] M. ASCE

Abstract

For seepage control features of embankment dams, design criteria and construction practice is reviewed as it is reflected in the experience of the U. S. Army Corps of Engineers. The generally accepted principles of embankment zoning are presented along with a discussion of some material property uncertainties that have been recognized following the unsatisfactory, in-place behavior of materials traditionally considered as being acceptable. Reliance is on the increase for internal filters, drains, and transition zones to perform a difficult variety of important functions ranging from the normal control of internal seepage, to effective defenses against deformation cracking and hydraulic fracturing. These multiple functions are identified and the need for more definitive specifications and thorough construction inspection is emphasized to assure effective translation of design intent to a satisfactory as-built condition. Recommendations are also offered for the enhancement of contract specifications to draw increased construction attention to filter zone placement techniques that will lessen the potential for segregated material concentrations as well as interzone intrusion beyond acceptable tolerances.

INTRODUCTION

Throughout the history of the U. S. Army Corps of Engineers, over 600 dams and navigation impoundment structures have been designed, constructed, and remain in operation today. With the oldest of these structures dating to 1884, many have since been built to provide impoundments that serve a wide variety of purposes. Of this total number, most consist either wholly or in part of earth or earth-rockfill components that have been designed and placed to function as water retention structures. The purpose of

---

[1]Chief, Geotechnical and Civil Branch, Engineering and Construction Directorate, Office of the Chief of Engineers, Washington, D. C.

31

this paper is to reflect upon some of the design and construction practices that have contributed to this depth of experience as compared to post impoundment operational performance. In so doing, some useful lessons learned can be identified that will be beneficial to the design and construction of similar structures in the future.

It is also of interest to note that during the four year period between 1977 and 1981, the Corps of Engineers was tasked under Public Law 92-367 to undertake a National Program of Inspection of Non-Federal Dams. The results of that nationwide effort (5) revealed that out of a total of over 8800 structures inspected, approximately 3000 were considered as having serious enough deficiencies to be classified as unsafe. And of this number, 842 or 28 percent included seepage as a matter of serious concern. The issue of seepage and leakage from surface impoundments therefore, is not only one in which the Corps of Engineers and similar agencies must contend, but is of such broader scope to be of interest and concern to the entire profession.

DESIGN PRACTICE

In the review of design practices for embankment seepage control measures, recognition must be given to the fact that the designer has essentially two major challenges with which he must contend. The first is the design of the embankment structure itself, where opportunity for control of fill properties is available in translating the design intent to the realities of construction. The second, and often more complex task is the proper adaptation of the embankment to its foundation, abutments, or adjacent structures, with all their complexities of geometry and variations of physical properties.

Success in achieving the first task is generally wide-spread. Embankment zonation principles which tailor volumes of borrow and required excavation to embankment requirements have been effectively applied, with proper attention given to relative permeability between embankment zones for internal control of phreatic surfaces. In addition, the designer's ability to predict each embankment zones' permeability characteristics has generally proven to be successful, through either laboratory analysis of remolded specimens of potential borrow or required excavation materials or analysis and field testing of prototype test fill embankments, or both, with opportunity for confirmation of such characteristics in the field during construction. Embankment zoning objectives representative in practice today can be summarized as follows:

-- To maximize the use of materials available from required project excavations and minimize the necessity for distant borrow,

-- To develop the least complex embankment cross
   section for economy of construction, with
   recognition given to logical, cost-effective,
   sequences of placement,

-- To provide an embankment zoning arrangement that
   properly defends against excessive seepage pressures,
   erosion, or piping both internally and at its contact
   with its foundation, abutments, or adjacent
   structures,

-- To provide an embankment with physical properties
   sufficient to remain safely stable under all anticipated reservoir operating conditions and natural
   hazards.

These considerations have contributed to many hundreds of
embankment dams which have successfully performed over many
years of service without any surface manifestation of
internal seepage control problems. Yet observations of
embankment surface cracking, as well as instrumental
evidence of internal deformations would lead one to conclude
that some internal cracking has undoubtedly occurred,
placing the burden of defense on built-in seepage control
features such as filters, transition zones, and drainage
blankets. Increasing emphasis and reliance is being placed
on such filter and transition zones both for internal
seepage control, as well as for defense against deformation
cracking or hydraulic fracturing of impervious embankment
components. The use of vertical or inclined "chimney
drains" and horizontal drainage blankets within embankment
cross sections is relied upon to provide effective interception of embankment seepage and its controlled conveyance
to the downstream toe. Downstream collection systems in the
form of controlled grade toe trenches, pipe trenches with
manhole inspection and intermittent flow measurement points,
and seepage collection/measurement ponds have routinely
been placed at most dams, and have furnished useful
performance data in relation to reservoir stages, when not
masked by tailwater or rainfall conditions. While it is
often a difficult task to separate seepage quantity
contributions of the embankment itself from that of its
foundation, abutments, or junctions with structures,
experience indicates that strictly internal embankment
seepage is comparatively minor. The success or failure of
such interception and collection systems has primarily been
a function of recognizing and interpreting the nature and
limits of underseepage flow regimes within natural
formations of foundations or abutments.

   Evidence appears to be building that some design
considerations or criteria formerly believed by many
practitioners to be entirely satisfactory, and supported by
many successfully performing embankments, now reveal some

areas of uncertainty and the need for refinement of some
traditional practices. While impervious core zones must be
composed of materials that are sufficiently impermeable to
maintain seepage losses through the embankment within limits
that can be safely controlled, and without compromise of
project purposes, mechanisms explaining poor resistance to
internal erosion and deformation behavior which can lead to
uncontrolled erosion have been identified. For example, the
erosion or piping resistance of proposed impervious embankment materials has routinely been recognized as a function
of plasticity, gradation, and achieved field compaction
(6) (9). However, examples have been reported where the
fine fraction of coarse, broadly graded materials have
undergone "selective" erosion and lacked sufficient
"internal stability" to resist significant and progressive
embankment damage (12). The coarse fractions of such
materials do not contain a sufficient distribution or mix of
grain sizes to prevent the internal migration and/or loss of
their fine fractions, even under relatively low gradients.
This question of "internal stability" has also been raised
in recent years as a possible area of concern for filter
materials as well, particularly those manufactured by
expedient, on-site crushing. Such materials often prove to
be gap-graded, lacking medium to coarse sand sizes, since a
more balanced grain size distribution either did not
naturally exist within the parent materials, or were not
producible. As discussed later, any segregation, or
construction process which allows it, only aggravates or
intensifies the problem further.

The dispersive tendency of some clay materials has also
been reported (8) (10) (11) and increased attention is being
given to the chemical interaction of soil constituents with
placement or reservoir water. In addition, the low tensile
strength characteristics of impervious core and other
proposed embankment materials is being given more thorough
recognition in positioning impervious zones within the cross
section, adjacent to structures, and against irregular
foundation/abutment surfaces for defense against deformation
cracking and hydraulic fracturing (1) (7).

The minimum width of impervious core zones also has
often been the subject of some debate. The practice that is
most generally followed establishes the minimum width at
approximately 25 percent of the maximum anticipated head
differential across the structure, and has resulted in
acceptable performance and manageable hydraulic gradients.
While a lesser dimension is theoretically acceptable, in
reality, the selected proportion is most often controlled by
the foundation conditions upon which the core zone is
placed, as predicted with difficulty by advance exploratory
investigations, and secondly, by the size of the core zone
and the available quantities of materials for its
construction.

The achieved density, in-place gradation after the rigors of field compaction, and in-place permeability are other important material properties of filter and transition zones that are and should be the subject of intense attention during design and construction. With respect to their seepage control function, the materials of such zones are heavily relied upon to: (1) intercept all through seepage emerging at downstream boundaries of impervious embankment components, including the contact of these features with foundation or abutment surfaces and with rigid structures, (2) possess sufficient hydraulic carrying capacity to safely convey intercepted seepage to points beyond the embankment limits, (3) be of such a gradation that will effectively prevent the migration of fines from protected embankment or foundation soils both with and without crack defects, (4) maintain permeability characteristics sufficient to allow the unimpeded transmission and passage of collected seepage, and (5) retain all of the above capabilities following the altering effects of field compaction or adverse seepage water chemistry. To successfully accomplish these many functions is indeed a formidable task. Nevertheless, the satisfaction of these requirements, particularly the filter requirements, has been attempted for many years through the application of traditional filter criteria, (3) (4) and has resulted in many successfully performing embankment dams. These criteria, which have evolved from laboratory tests and field experience, apply to cohesionless or slightly cohesive soils with approximately parallel grain size disbribution curves, and specify that the gradation of proposed filters meet two basic requirements. First, for stability against the migration of the protected base soil into the filter, the

$$\frac{15\% \text{ size of filter material}}{85\% \text{ size of material being drained}} = \frac{D_{15F}}{D_{85B}} < 5 \ldots (1)$$

and the

$$\frac{50\% \text{ size of filter material}}{50\% \text{ size of material being drained}} = \frac{D_{50F}}{D_{50B}} < 25 \ldots (2)$$

Secondly, to insure that the filter material is sufficiently more permeable than the base material being drained, the

$$\frac{15\% \text{ size of filter material}}{15\% \text{ size of material being drained}} = \frac{D_{15F}}{D_{15B}} > 5 \ldots (3)$$

For filter and protected base materials with grain size distribution curves that are not approximately parallel, or for proposed filter materials composed primarily of crushed stone, some adjustment of the ratios given above may be required, and should be checked by laboratory filtration tests.

CONSTRUCTION PRACTICE

   While the factors discussed above are certainly of
importance in the design of seepage control measures within
embankments, experience has shown that of equal, if not
greater importance is the necessity to properly specify,
inspect, and assure that the actual construction process
translates the design intent into a satisfactory "as built"
condition. Judgments are often made based upon clinically
idealized laboratory specimens, definitive lines or
boundaries are placed on construction drawings often without
regard to the practical reality of construction methods, and
the equipment necessary for high volume placement.

   With respect to an important internal embankment filter
zone for example, many construction specifications deal with
placement of these materials in a similar fashion; i.e.,
gradation requirements to be met in-place after compaction,
the necessary loose placement layer thickness, appropriate
moisture control requirements, and the specification of
either the desired end product density to be achieved or the
number of compaction coverages to be applied to each layer
by specified types of equipment. What is most often not
specified is guiding language dealing with the placement and
spreading procedures or techniques that are to be used and
the interzone discipline that is to be followed as the
embankment construction proceeds in increasing increments of
elevation. These points are often left to the discretion of
the contractor,and can result in unsatisfactory concentrations of severely segregated materials or uncompacted
zone boundaries,which can have a very significant bearing on
the ultimate performance of the embankment, its deformation
behavior, and its internal seepage control measures after it
is placed in service. Figures (1) through (3) illustrate a
variety of placement techniques that have been followed
depending upon embankment zone inclination and specification
controls on elevation limits between adjacent zones as
embankment construction proceeds.

   Depending upon actually specified lift thicknesses,
specified tolerances permissible at the boundaries between
adjacent zones, as well as field interpretation, large
differences can result in placed filter zone widths, as can
disputes on actual construction quantities. While Figures
(1) through (3) illustrate somewhat idealized and consistent
interzone boundaries, it should be recognized that such
variations can be even greater depending upon the placement,
spreading, and compaction techniques followed during actual
construction. Oftentimes, contractor preferred high volume
bottom discharge dumping can and has shown intrusion and
interfingering between adjacent embankment zones far greater
than illustrated. Use of spreader boxes and front end
loader equipment on the other hand, has achieved a lesser
degree of intrusion, but at the expense of production rate.

Shallow, 1-2 lift high formwork has also been used between zones as an extreme example of maximum control, but would only be of practical and economical value on low volume, specialized applications. The construction procedure portrayed in Figure 1 is a popular approach preferred and taken by many embankment engineers, and results in filter zone widths that are often greater than specified in contract documents, and can also result in quantity overruns on large projects, and unless carefully controlled, the likelihood of greater intrusion into adjacent embankment zones.

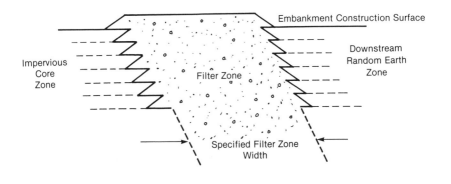

Figure 1. Embankment Construction to Uniform Elevation Filter Zone Leading

The placement conditions shown in Figure 2 have also occurred on several occasions, have resulted in effective filter zone widths generally less than required, and were caused by misinterpretation in the field.

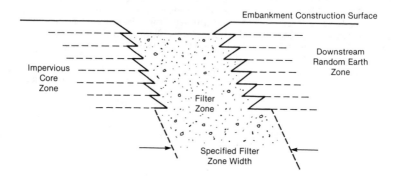

Figure 2. Embankment Construction to Uniform Elevation Filter Zone Following

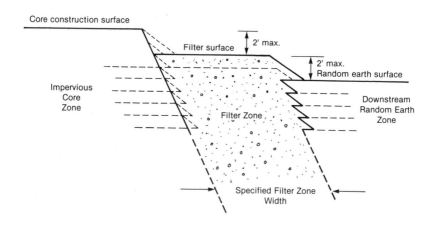

Figure 3. Embankment Construction Limiting Zone Elevation Differences

Figure 3 illustrates another technique popular among design
engineers where the impervious core zone is maintained
higher than adjacent zones throughout construction, and each
adjacent zone is required to be stepped in increments not
exceeding one to two lifts, typically two feet (0.6 m)
maximum. Where more positive boundary control is desired,
shoulder trimming of the higher zone can be specified prior
to placement of the adjacent lower filter zone, but can
enhance contamination unless carefully done. Similarly, a
trenching approach has occasionally been used after earth
zone materials have been placed and compacted, to re-
establish contact with the buried surface of the filter zone
for continued filter zone placement within the confined,
neat boundaries of the excavated trench. Use of this
procedure also can result in both contamination and
loosening of placed materials at depth unless undertaken
carefully.

While the minimum width of the filter zone that is also
intended to act as a drain or interceptor of through-
seepage can be calculated to provide sufficient hydraulic
carrying capacity (2), practical minimum construction widths
of 10-12 feet (3.1 m - 3.7 m) most often control the
dimension selected for use, and are significantly greater
than theoretically necessary. Such greater widths, unless
restricted by extreme economical considerations, also
contain sufficient excess to accommodate some limited
interzone contamination.

In addition to traditional specification requirements,
other important factors to consider in preparing contract
document provisions include:

-- The overall technique required to be followed in
   constructing the filter zone in relation to adjacent
   embankment zones for both space restricted and
   unrestricted regions of the embankment,

-- The moisture conditions and spreading requirements
   that will minimize segregation and interzone
   intrusion beyond specified tolerances,

-- The requirements for deliberate compaction of zone
   boundaries for both space restricted and unrestricted
   embankment areas,

-- A specification requirement for the random staggering
   of leading edges of embankment zone placement lanes,
   particularly at zone boundaries, so that segregated
   loose accumulations of embankment materials are not
   continuous along any plane across the embankment
   cross section including abutment contacts.

It can generally be stated that embankment seepage control design concepts as discussed earlier have met with widespread success from the standpoint of post impoundment behavior, especially when compared with one's predictive ability to evaluate the complexities of in situ foundation/ abutment conditions. But these successes have not occurred without the need for careful inspection during construction and adherence to the well accepted principle that the design is not complete until construction has been accomplished. Such close inspection is essential to insure that the design intent is fulfilled and that changes in delivered material characteristics or conditions exposed, however subtle and insignificant they may seem, do not compromise the fundamental requirements envisioned in the original design. The importance of such field inspection and surveillance is illustrated on Figure (4).

The design and construction contract documents for the project illustrated called for an 85 feet (26 m) high embankment wing dam to be constructed adjacent to and against a concrete gravity structure. As indicated, the embankment cross section consisted of a vertically faced, approximately 28 feet (8.5 m) wide, centrally located impervious core, flanked upstream by a six foot (1.8 m) wide vertical filter zone and external pervious shell, and down- stream, by a zone of silty and clayey sand random earth. Following construction, and as initial impoundment was nearly completed, an emergency seepage condition developed. The pool was lowered, and subsequent investigative measures taken to determine the cause(s). Investigative excavation and mapping of the completed embankment was carefully undertaken, and the actually placed embankment zone boundaries compared to the specified embankment zoning was revealed as shown on Figure 4. Actually placed zone boundary variations and intrusion of upstream filter materials within the intended impervious core zone as great as 14 feet (4.3 m) were discovered, and, in combination with intrusions from the downstream random zone, effectively reduced the width of the core by as much as 50-60 percent. While other factors were considered as the primary contri- buting causes of the emergency seepage event, the interzone embankment boundary conditions revealed in the investigative excavations nevertheless, illustrate the need for increased attention to be devoted to specification coverage of the subject as well as the importance of persistent inspection during construction. Had the embankment and impoundment levels been higher, and had a less conservative zoning configuration been adopted, a potential for more severe consequences would have existed.

SUMMARY

Traditional approaches to embankment design in terms of zoning configuration practice, the relative permeability of

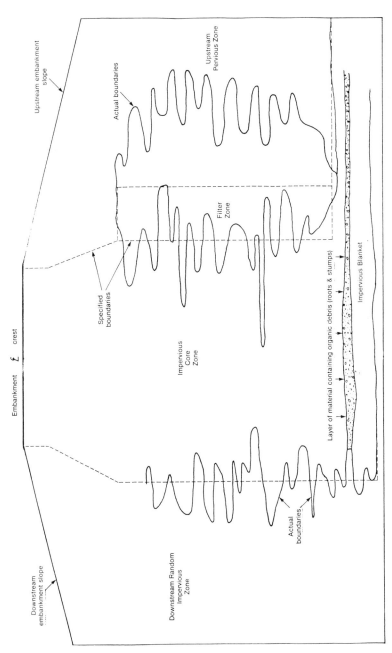

Figure 4. Typical Embankment Cross-Section

adjacent zone materials, the use of intercepting drains and
blanket drain outlets, and filter and transition zones, are
all well established design considerations that have clearly
contributed to large numbers of successfully performing
embankment dams. However, throughout long periods of
service, surface and instrumental expressions often suggest
that unpredicted distress is indeed occurring, placing
increased reliance upon defensive filter, drain, and
transition zone features, as well as the foresight and
predictive ability of design personnel for future designs.
Careful and frequent surface observations coupled with the
profession's relatively recent ability to instrumentally
monitor and quantify internal deformations, have led to an
increased understanding of embankment multidimensional
behavior, both short and long-term. Even with such enhanced
capability, dependence upon defensive seepage control
features of embankments remains undiminished.

As prudent design efforts continue, the importance of
more definitive construction specifications and construction
surveillance/inspection by design personnel cannot be over-
emphasized. Proper embankment construction and the
translation of design intent to an effective as-built
condition is a process that is far more intricate and
demanding than any other form of earthwork construction.
The language of contract documents and the degree of
construction control/supervision therefore, must be thorough
and convincing.

Of all the seepage control considerations required for
sound embankment design, mention was given earlier that
designers are essentially faced with two major challenges;
the first being the design of the embankment itself, and the
second, often more complex task of adapting the embankment
to its foundation, abutments, or adjacent structures. There
is little doubt that this latter task is the more complex
and requires a significant commitment of interpretive skills
and thorough subsurface investigation. In the Corps of
Engineers experience, as perhaps for others as well, such
foundation seepage conditions which develop either after
initial impoundment, or following record pool levels after
many years of service, continue to place the greatest
demands upon operational and maintenance resources.

APPENDIX I - REFERENCES

1.   Al-Hussaini, M. and Townsend, F C., "Investigations of
      Tensile Strength of Compacted Soils," Miscellaneous
      Paper S-74-10, June 1974, U. S. Army Waterways
      Experiment Station, Corps of Engineers, Vicksburg,
      Miss.

2.   Cedergren, H. R., Seepage, Drainage, and Flow Nets, 2nd
      Edition, 1977, John Wiley & Sons, New York.

3. Department of the Army, Office of the Chief of
    Engineers, "Soil Mechanics Design - Seepage Control,"
    EM 1110-2-1901, February 1952 (under revision),
    Washington, D. C.

4. Department of the Army, Office of the Chief of
    Engineers, "Earth and Rockfill Dams - General Design
    and Construction Considerations," EM 1110-2-2300, May
    1982, Washington, D. C.

5. Department of the Army, Office of the Chief of
    Engineers, Final Report to Congress, National Program
    of Inspection of Non-Federal Dams, May 1982,
    Washington, D. C.

6. Gibbs, H. J., "A Study of Erosion and Tractive Force
    Characteristics in Relation to Soil Mechanics
    Properties," Report No. EM-643, Febraury 1962, U. S.
    Department of the Interior, Bureau of Reclamation,
    Denver, Colo.

7. Nobari, E. A., Lee, K. L., and Duncan, J. M.,
    "Hydraulic Fracturing in Zoned Earth and Rockfill
    Dams," Contract Report S-73-2, January 1973, U. S. Army
    Waterways Experiment Station, Corps of Engineers,
    Vicksburg, Miss.

8. Perry, E. B., "Piping in Earth Dams Constructed of
    Dispersive Clay; Literature Review and Design of
    Laboratory Tests, "Technical Report S-75-15, November
    1975, U. S. Army Waterways Experiment Station, Corps of
    Engineers, Vicksburg, Miss.

9. Sherard, J. L., "Influence of Soil Properties and
    Construction Methods on the Performance of Homogeneous
    Earth Dams," Sc.D Thesis, March 1952, Harvard
    University, Cambridge, Mass. (Also published as U. S.
    Department of the Interior, Bureau of Reclamation,
    Technical Memorandum 645, January 1953).

10. Sherard, J. L., Decker, R. S., and Ryker, N. L.,
    "Piping in Earth Dams of Dispersive Clay," Proceedings
    of the Specialty Conference on Performance of Earth and
    Earth-Supported Structures, ASCE, June 1972.

11. Sherard, J. L. and Decker, R. L., "Summary -
    Evaluation of Symposium on Dispersive Clays,"
    Dispersive Clays, Related Piping, and Erosion in
    Geotechnical Projects, American Society for Testing and
    Materials, Special Technical Publication No. 623, May
    1977.

12. Sherard, J. L., "Sinkholes in Dams of Coarse Broadly
    Graded Soils," 13th ICOLD Congress, India, Vol II,
    1979.

EARTH DAM SEEPAGE CONTROL, SCS EXPERIENCE
by James R. Talbot [1] and David C. Ralston [2], Members, ASCE

ABSTRACT

The U.S. Department of Agriculture's Soil Conservation Service (SCS) designs and administers construction of a large number of small- or medium-sized dams. SCS experience has shown that nearly all dams crack to some degree, regardless of size. Cracking can lead to concentrated seepage, erosion, and eventual failure, particularly in dams constructed of dispersive clays, certain broadly graded soils, and other erodible soils. Through laboratory tests and full-scale field tests, SCS has studied the use of sand and gravel filters to seal cracks or other concentrated leaks in dams. Cracked homogeneous dams having no filter zones have been repaired by trenching into the dams to facilitate filter installation. The trend in SCS is toward the use of filters instead of structural walls (anti-seep collars) or other measures to intercept seepage and prevent erosion. SCS engineers have developed laboratory procedures for determining the suitable particle size distribution of filters for specific soils used in dams.

Introduction

A wide variety of seepage control measures are used in dams designed by the U. S. Department of Agriculture's Soil Conservation Service (SCS). These dams vary widely in purpose and size, from very small farm ponds in remote areas to large dams with populated areas downstream.

SCS experience shows that some cracking occurs in small as well as large dams. In those dams constructed of dispersive clays, certain broadly graded soils, and other erodible soils, seepage flow in cracks has caused erosion.

This paper describes SCS's experience in dam construction and drainage design and examines case histories of cracking in dams constructed of erodible soils. It focuses on SCS evaluations of sand and gravel filters used to prevent excessive seepage and erosion through cracks.

Size, Type, and Number of SCS Dams

SCS designs and directs the construction of many dams annually. The dams range from only a few feet to over 100 ft (30 m) in height and vary in type from small concrete spillways contained within a short embankment to earth or earth-rock embankments containing 1 million cubic yards (760,000 m³) or more of earth fill with appurtenant

---

[1] National Soil Engineer, USDA, Soil Conservation Service, Wash., D.C.

[2] National Design Engineer, USDA, Soil Conservation Service, Wash., D.C.

spillways. The purposes for which the dams are built include
floodwater retarding, water storage for irrigation and recreation,
gully control and grade stabilization, sediment control, agriculture
waste storage, fish ponds, and storm water management.

The dams for which SCS provides technical assistance are for soil
and water conservation districts, flood control districts, irrigation
districts, county and city governments or other local groups, as well
as individual farmers and ranchers. These individuals or groups own
the dams, and after construction, operate and maintain them. Under
some operating authorities, a planned agreement for operation and maintenance is drawn up between the owner and SCS.

Most of the dams built with SCS assistance are in remote areas and are
so small that they present no potential hazard to life or to important
utilities in the event of failure. Some dams, however, are located
where, if they failed, there would likely be loss of life or serious
damage to important buildings or utilities. SCS uses a classification
system to indicate the degree of potential hazard as follows:

   Class (a).--Dams located in rural or agricultural areas where
   failure may damage farm buildings, agricultural land, or
   township and country roads.

   Class (b).--Dams located in predominantly rural or agricultural
   areas where failure may damage isolated homes, main highways, or
   minor railroads or interrupt the use or servicing of relatively
   important public utilities.

   Class (c).--Dams located where failure may cause loss of life
   and serious damage to homes, industrial and commercial buildings,
   important public utilities, main highways, or railroads.

Table 1 shows the number of SCS dams in each classification
according to an estimate made in December 1983.

TABLE 1.--ESTIMATED NUMBER OF SCS DAMS, BY CLASS, DECEMBER 1983

Classification	Number of Dams [1]	% of Total
(a) Low Hazard	17,232	67
(b) Moderate Hazard	1,905	7
(c) High Hazard	1,638	6
Unverified [2]	5,111	20
Total	25,886	

---

[1] Significant dams, including all class c and b dams and those class
a dams having a height of more than 6 ft (1.8 m) and a storage capacity
of 50 acre-ft (61,700 m^3) or more, or height of more than 25 ft (8 m)
and a storage capacity of more than 15 acre ft (18,500 m^3).

[2] These dams have not had recent field verification by SCS, but most
are expected to be in the low hazard category.

Preliminary data from a 1984 inventory of significant SCS dams show
that 87% of the dams are less than 40 ft (12 m) high, 10% are 40
to 60 ft (12 to 18 m), 2% (454) are 60 to 100 ft (18 to 30 m), and
about 1% (56) are 100 ft (30 m) to more than 150 ft (46 m). For 84%
of the dams, the embankment volume is less than 100,000 yd^3
(76,000 m^3); for 11% the volume is 100,000 to 1 million yd^3
(76,000 to 760,000 m^3); for less than 1% (36) it is over 1 million yd^3
(760,000 m^3); and for the rest there is no record. Reservoir storage
volume is less than 10,000 acre-ft (1.23 x 10^6 m^3) for 97% of the SCS
dams; 10,000 to 25,000 acre-ft (1.23 X 10^6 to 30.8 X 10^6 m^3) for 1%;
and more than 25,000 acre-ft (30.8 X 10^6 m^3) for less than 1%. For the
remaining 1 percent there is no record of storage volume.

Of the dams inventoried in 1984, 25% were built prior to 1960; 43%,
between 1960 and 1970; 29%, between 1970 and 1980; and 3%, since 1980.

The dams built with SCS assistance, with rare exception, are earth or
earth-rock embankments having a principal spillway consisting of a
pipe conduit through the embankment foundation and a tower (riser)
type inlet. A high-level (emergency) spillway is usually excavated
into the abutment materials. These earth spillways are usually
vegetated earth or rock. A small percentage have structural (reinforced concrete) emergency spillways.

Foundations of these dams commonly are yielding alluvial soil.
Natural abutment slopes are usually 3 horizontal to 1 vertical or
flatter and consist of glacial or residual soils. Some abutments are
steep and some have rock at or near the surface. Foundation preparation consists of stripping the surface organic layer, removing soft,
low-density soil layers or pockets, and excavating a shallow cutoff
trench to a relatively impervious soil layer or bedrock if feasible.
For a small percentage of the structures, grouting is needed to
reduce seepage through weathered or jointed rock zones.

The smaller dams are usually homogeneous embankments of fine-grained
soil from the emergency spillway excavation and from the reservoir
bottom. Additional soil, if needed, is often obtained from the
abutments in the reservoir area. If coarse-grained soil is available, a zoned embankment is often built by selective placement of
the coarse materials in the outer shells. Durable rock from the
spillway excavation is used for upstream slope protection at the
normal reservoir water line and sometimes for rock shell zones or to
stabilize outlet channels; however, zoned or rockfill dams constitute
a small percentage of SCS dams.

The principal spillway is usually pipe placed on or in the foundation
soils near the embankment base. The inlet is sized to permit use of
maximum conduit capacity under pressure flow conditions. Over the
years, the traditional structural cutoff or anti-seep collars have
been constructed at an even spacing along principal spillway conduits
and one or two locations on open spillways. These were used to
extend the seepage path along the contact between the smooth
structure and the embankment soil and intercept zones of cracking.

Drainage Designs for SCS Dams

Because dams built with SCS assistance have a wide variety of
purposes, a wide variety of seepage control measures are used. Early
designs of small earthfill dams used very little drainage. Many
flood control dams built by SCS have homogeneous cross sections with
small foundation trench drains under the downstream sections of the
dams. Except for a very few cases, these have been successful and
have served for many years.

Except during flooding, most flood control dams contain water at a low
level for sediment accumulation and incidental recreation. Some of
these dams may have an embankment drain that extends up to the level
of the low pool. Many of the smaller dams of this type are homogeneous
with a downstream foundation drain as the only drainage system.

Irrigation storage structures, water regulating reservoirs, and large
dams that remain full for long periods usually have extensive drainage
systems. This is also true for other dams with geologic conditions
or embankment soils having the potential for large seepage rates or
piping. These systems may include vertical embankment drains extending
to the top of the dam, abutment drains, foundation blanket drains
and/or trench drains. On occasion, relief wells are incorporated to
intercept confined permeable layers or broken rock zones in the
foundation in small as well as large dams. In all cases the drainage
system is designed in conjunction with cutoff measures to intercept
water that seeps beyond cutoff trenches, grout curtains, sheet pile
walls, or other measures. Drains are also placed to intercept zones
of potential cracking. Although these drainage systems have been
used on some SCS dams, a large percentage of the dams have very
little potential for seepage problems and were designed with little
or no drainage. In recent years, SCS has placed more emphasis on
intercepting potential zones of cracking with filters because evidence
shows filters and drainage to be the best solution for this problem.

Evidence of Cracking

SCS experience and the documented experiences of others (7) have
caused us to conclude that some cracking occurs in almost every dam
(6). This conclusion is verified by SCS experience with failures or
other incidents involving small dams constructed of dispersive clay
soils and other highly erodible materials. A few of these structures
having no internal filters failed or experienced piping or tunnel
erosion on the first filling after water under very low head had been
in the reservoir only a few hours. Tunnel erosion has also occurred
on or near the surface of many dispersive clay dams. Laboratory
studies and field performance have shown that dispersive clays will
not erode or cause erosion tunnels unless there is an opening, crack,
or other flow channel larger than the normal pore (void) spaces in
the soil (12, 14, 18). In all these cases, the water reached the
discharge point on the downstream slope much too fast for passing
through the normal pore spaces. Cracks or other openings were either
found or suspected by investigating teams who studied the cause of
the incident.

Many dams have cracked because of hydraulic fracturing through a path
in the embankment along which the lateral stress has been reduced by
stress transfer from differential settlement. This problem is so
common because most dams for flood control or most small reservoirs are
located on alluvial fans or intermittent water courses. These areas
usually contain deep, compressible alluvial or eolian soils, irregular
profiles, buried channels, and areas of contrasting soil properties.
The economics of deep soil removal from the foundation is prohibitive
for structures of this size, so the dams are usually constructed
on these compressible foundations. Lateral stress reduction from
differential settlement is the usual condition. Sometimes the
settlement is sufficient to create tension zones and cracking before
any water reaches the zone. Other causes of cracking are desiccation
and nonuniform or inadequate compaction in the embankment or around
structures.

A small percentage of dams exhibit problems from cracking. In most
cases, the cracking is likely minor and confined to abutment areas,
areas near structures such as outlet conduits, or areas above
nonuniform foundation surfaces or soil conditions. In normal
(nondispersive) clay soils with some resistance to piping, the cracks
likely swell shut or fill up by slaking rather quickly so that no
leakage appears or is noticed. For dams with properly designed
filter-drainage zones, any leaks through cracks or other openings are
immediately sealed at the filter face, or the seepage velocity is
reduced by partial sealing and the crack fills by slaking as water
penetrates the sides of the crack. No leakage is detected and no
problem develops so there is no awareness of the cracking.

Case Histories of Cracking in SCS Dams

Dispersive clay dams. In Oklahoma, Mississippi, and other locations,
several small SCS dams constructed of dispersive clay have failed or
shown excessive tunnel erosion (2,6,8,16). Other structures such as
channels in dispersive clay have also experienced erosion problems.
In each case the erosion apparently started in a crack or other
opening. In the cases of tunnel erosion near the surface, the cracks
were caused mostly by desiccation and were usually about 3 ft (1 m)
deep and spaced 6 to 20 ft (2 to 6 m) apart.

Several successful treatment measures have been used on these structures. Dunnigan, Sherard, Decker, Ryker, Forsythe, and others have
documented the successful use of lime and alum to treat dispersive
clays (2,5,6,15). Gypsum has also been successful as a chemical
treatment to prevent dispersion. SCS has found that placing a layer
of sandy gravel over the surface stops tunnel erosion near the
surface. This was tried on an eroded channel bank in Oklahoma. The
erosion was repaired and a thin blanket of sandy gravel was placed
over the surface. Despite cracks observed in the channel bank before
the gravel blanket was applied, no further problems have occurred.
Two processes are likely involved in preventing these problems: (a) the
gravel blanket prevents some of the desiccation cracking and (b) the
filter action of the sandy gravel prevents the movement of soil from
the existing cracks. The use of filters to prevent erosion of

dispersive clays has been further verified by laboratory studies and
observations of field performance (12, 14).

Cracked dams in arid regions. Many SCS dams are built in the arid
southwestern United States to protect agricultural land as well as
rural and urban areas from flooding. These dams are often built on
intermittent water courses or across alluvial fans, where floods
occur only in the event of high intensity rain storms. The reservoirs behind these dams are dry except during the short and infre-
quent periods of flooding. After each storm they drain rapidly
(within 3 to 10 days).

Many of these dams have been built in Arizona over the past 30 years.
For the most part they are long, low dams with homogeneous cross
sections and no drainage zones. Typically the dams are 20 to 30 ft
(7 to 10 m) high and up to 5 miles (8 km) long. Many of the dams
have experienced cracking that appears to be associated with desiccation (19). The depth of drying is approximately 9 to 12 ft (3 or
4 m). The cracks are generally perpendicular to the centerline, are
tapered from ¼ inch to 2 inches (6 to 50 mm) wide at the top to
hairline at depth, and extend to the depth of drying. A few cracks
are deeper and may extend to the foundation, indicating the possibility of foundation settlement (19, 20).

The cracks in these dams were discovered during scheduled inspec-
tions. None of these dams failed or eroded from seepage through
cracks; however, most of them had stored either no water or only
small floods. One cracked dam stored a large flood that completely
filled the reservoir for a period of 3 weeks, yet no problem
developed. The dams are constructed of erodible soils and are
located above populated areas where the hazard from failure is high.
SCS engineers considered these dams to have a high potential for
failure from erosion due to concentrated seepage in the open cracks.

In 1978, SCS began repairing cracked dams in Arizona using a filter-
drainage system to seal the cracks. This procedure consists of
trenching to the depth of cracking and filling the trench with a
properly graded gravelly sand filter as shown in Fig. 1 (19,20). The
design includes drain outlets, spaced at intervals generally less
than 100 ft (30 m), that incorporate drainfill coarser than the
filter. When the reservoir fills with water and seepage begins
flowing in cracks, the filter traps any moving soil particles at the
soil-filter interface and thus seals the crack. A total of 6 dams
were repaired in Arizona by the end of 1984 using this method.

Dams on deep loess and other collapsible soils. Small dams built on
collapsible soils in the deep loess area of the Midwest and in some
alluvial outwash areas of the arid Southwest have experienced major
cracking. The cracking occurs after the foundation is wetted from
storage of flood water. The wetting causes the soil structure to
collapse and the foundation to settle rapidly, usually more than
1 ft (0.3 m). These rapid settlements almost always cause severe
cracking.

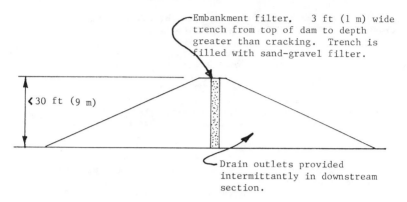

Embankment filter. 3 ft (1 m) wide
trench from top of dam to depth
greater than cracking. Trench is
filled with sand-gravel filter.

‹ 30 ft (9 m)

Drain outlets provided
intermittantly in downstream
section.

FIG. 1.   REPAIR MEASURE FOR CRACKED DAMS LESS THAN 30 FT. (9 m)
HIGH USING SAND-GRAVEL FILTERS

SCS has adopted procedures for testing soil conditions and for
removing the problem soils or treating the foundation area to consol-
idate the soil before completing the construction of the dam.
Several dams constructed over 20 years ago, before these removal and
treatment procedures were adopted, have cracked severely.   Some of
these dams have been repaired using filter-drainage systems similar
to those used for the Arizona dams.   Dams under 30 ft (9 m) in height
are trenched from the top using a deep trenching machine to intercept
all the cracks.   The trench is filled with filter material as shown
in Fig. 1.   Dams over 30 feet (9 m) high require a combination of
excavation and trenching to install a filter, as shown in Fig. 2.

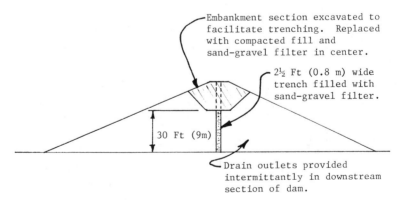

Embankment section excavated to
facilitate trenching. Replaced
with compacted fill and
sand-gravel filter in center.

2½ Ft (0.8 m) wide
trench filled with
sand-gravel filter.

30 Ft (9m)

Drain outlets provided
intermittantly in downstream
section of dam.

FIG. 2.   REPAIR MEASURE FOR CRACKED DAMS OVER 30 FT (9 m) HIGH
USING SAND-GRAVEL FILTERS.

Dams constructed of broadly graded soils. Sinkhole development from
erosion of certain broadly graded soils has occurred in several dams
throughout the world (9). A similar incident occurred during the
spring of 1984 at Brodhead Dam, a dam designed by SCS and constructed
of glacial soils in the northeastern United States. The dam is an
earthfill structure 88 ft (27 m) high, constructed for flood control
with no permanent storage, (the reservoir is dry except during
flooding). At the time the incident occurred, the dam had been in
service for 9 years and had experienced one or two low-level fillings
each year. In 1984, a large flood filled the reservoir to the upper
spillway crest and maintained water in the reservoir for about 10
days. After the reservoir was empty a large sinkhole (cavity) was
found near the crest on the downstream side. Excavations into the
dam exposed numerous large cavities including vertical holes as
much as 8 ft (2.5 m) in diameter and 28 ft (8.5 m) deep. One main
horizontal cavity about 4 to 6 ft (1.2 to 1.8 m) in diameter connecting the vertical holes was also discovered. Some pockets of coarse
gravel soils were found near the lower elevations of the cavities.
The cavities appeared to terminate downstream at the blanket or
embankment drain. Fig. 3 shows the sinkhole locations.

The dam was basically a homogeneous cross section constructed of a
broadly graded soil of glacial origin. The embankment material was
classified as a silty sand (SM) with gravel. A plot of the grain
size distribution curve is shown in Fig. 4. The drainage system
included an embankment (chimney) drain extending about 15 ft (4.5 m)
up into the dam above the foundation level and down into the foundation on the back side of the cutoff trench. A blanket drain over the
lower left abutment covered an area of potential seepage through a
weathered rock zone. The drain consisted of uniform (narrowly
graded) gravel and was designed according to the typical criteria
utilizing the D_{15} size of drainfill as 5 times the d_{85} size of the
embankment soil. (See Appendix II for notation.) A plot of the
drainfill gradation is shown in Fig. 4.

Brodhead Dam is located a short distance upstream from a small
community, most of which would be inundated in the event of a sudden
embankment failure with a full reservoir. The dam was considered
unsafe in the event of a flood large enough to fill the reservoir.
SCS, therefore, breached the dam to remove the hazard by excavating
the earthfill adjacent to the left abutment where the cavities were
found. The area containing the cavities was inspected and the
abutment exposed during the breaching process.

The left abutment consisted of weathered shale bedrock. The surface
was on about a 1 (horizontal) to 1 (vertical) slope with some areas
near vertical. The left abutment rock at the dam centerline appeared
to be tight and massive; however, the rock upstream and downstream
from the centerline was highly weathered with many open joints and
fractures. The weathered zone was 5 to 10 ft (1.5 to 3.0 m) thick
(perpendicular to the slope) and consisted of thinly bedded and
jointed shale with many open joints. The earthfill adjacent to the
weathered shale on the upstream side was wet for a distance of 1 to
1-1/2 ft (0.3 to 0.5 m) from the rock surface into the fill. There
was also evidence of water in the joints of the weathered rock and

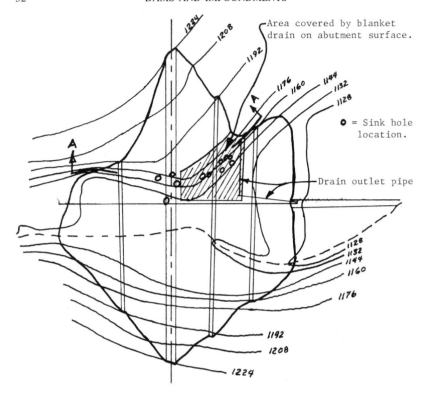

Area covered by blanket drain on abutment surface.

O = Sink hole location.

Drain outlet pipe

PLAN SHOWING SINK HOLE LOCATION

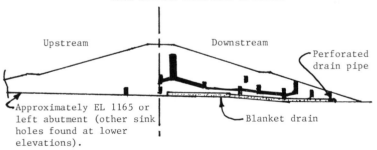

Upstream   Downstream

Perforated drain pipe

Approximately EL 1165 or left abutment (other sink holes found at lower elevations).

Blanket drain

SECTION A-A; APPROXIMATE PROFILE OF CAVITIES

FIG. 3.   BRODHEAD DAM SINK HOLE LOCATION

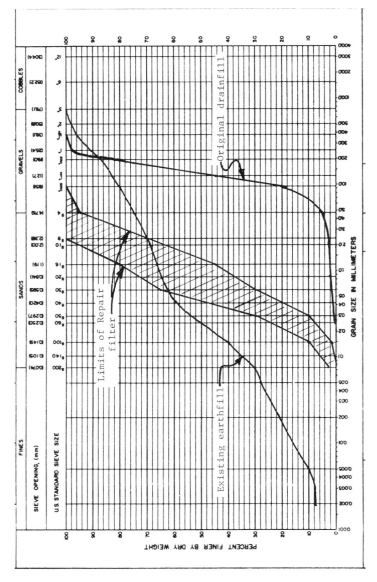

FIG. 4. BRODHEAD DAM GRAIN SIZE DISTRIBUTION CURVES FOR EARTHFILL, ORIGINAL DRAINFILL AND REPAIR FILTER

the appearance that some flow channels had been created through the
numerous joints of the weathered rock.

The upstream and downstream areas were searched for an indication of
inlets or outlets to concentrated leaks. None were found. There was
no observable delta or deposit of the material missing from the
embankment cavities even though the volume of the cavities was
estimated to be about 250 yd^3 (190 m^3).

A literature search on the problem revealed a study by Sherard
reporting a number of incidents in the past 20 years in which sinkholes or craters have appeared in the crest or slopes of dams (9).
Each of these dams was constructed of a broadly graded soil, usually
of glacial origin, ranging from a maximum particle size of about 3
in. (75 mm) down to about 15 to 30% passing the No. 200 sieve. For
such soil the particle size distribution curve on the normal
semilogarithmic plot is a straight line or is slightly concave
downward. The fines in these soils are either nonplastic or of low
to medium plasticity.

Glacial tills and other broadly graded soils are often thought to be
relatively well graded because all the particle sizes are present in
approximately equal quantities. Well graded soils are usually
considered self-healing when used in a dam. That is, they are
resistant to erosion and will adjust to fill any opening or leak
that may develop. The problem is, the soil at this site with a
straight line plot, unlike most soils is not well graded and, in
fact, has such a wide range of soil particles that the fine portion
can move through the coarse portion with seepage water movement. A
stable, well graded soil has a particle size distribution curve which
is sharply concave upward when plotted on the semilogarithmic chart.
In contrast, the straight line plot of these broadly graded soils
indicates a deficiency in the medium and coarse sand, so that voids
between the gravel particles are filled with fines and fine sand
particles, which can be moved between the gravel particles by
flowing seepage water. The finer portion is washed away leaving the
gravel in a cavity. Sherard (9) has shown that if these soils were
divided at some sieve size, the coarse portion does not meet the
proper filter gradation requirements for the fine portion (i.e., the
D_{15} of the coarse portion is more than 5 times the d_{85} of the fine
portion). Dividing the sample into two parts can be accomplished
mathematically at any particle size from a grain size distribution
curve of the entire soil. The D_{15}/d_{85} ratio can be easily determined for any dividing point along the curve.

A likely process by which the cavities developed in the Brodhead Dam
is as follows:

1.   Water entered the weathered zone of bedrock upstream of the toe
      on the left abutment.

2.   Seepage traveled through the many cracks and open joints of the
      rock and accumulated under pressure a short distance upstream
      from the embankment centerline. The weathered rock had been
      removed from the point of the protruding left abutment near

centerline in a cutoff trench. The seepage water pressure at
the upstream side of the narrow cutoff trench was likely equal
to the full reservoir pressure. This forced seepage through the
earthfill adjacent to the steep rock abutment.

3. High gradients caused hydraulic fracturing or the steep abutment
    caused differential settlement cracking in the fill adjacent to
    the rock abutment.

4. Water running under pressure through small cracks eroded the
    fine portion of the broadly graded fill either into the coarse
    blanket drain or into the downstream weathered rock zone. Once
    in suspension, these fines may have passed through the coarse
    drainfill or the rock abutment and on downstream with the
    seepage water. They were observed to have been deposited in the
    downstream coarse blanket drainfill near the base of the abutment
    and below the elevation of the outlet collector pipe. Some were
    likely carried on downstream through the drain outlet without
    making a visible deposit.

5. The sinkholes were formed by progressive caving of a complex
    system of cavities created by the selective erosion of the finer
    portion of the broadly graded earthfill materials. The coarse
    portion was left as a deposit in the bottom of the cavities.

Laboratory tests made by SCS consisted of placing the embankment soils
in a test cell over a coarse filter having a gradation similar to the
one designed for the drainage zones in this dam. These tests were
similar to those reported by Sherard, et al. (14). Water under pressure
was run through the soil and filter. When water pressure was increased
to the point that hydraulic fracturing occurred, the fine portion of
the embankment soil eroded into and through the coarse filter and a
large volume of the fine portion was lost. Additional tests verified
that a finer filter gradation selected in accord with the results of
an SCS laboratory filter study immediately seals cracks and prevents
erosion. Repair of the dam was initiated in the fall of 1984. An
embankment drain will be installed using a properly graded sand
filter verified by tests to seal cracks in the embankment material.
The filter gradation used in repair is shown in Fig. 4.

Studies on Sealing Cracks in Embankment Dams

SCS has conducted laboratory and field studies to evaluate the use of
sand or sand and gravel filters for sealing concentrated leaks in
embankment dams. Sherard et al. (13, 14) have published some results
on the laboratory work. Other papers on laboratory and field work
are in preparation.

The laboratory studies entailed running water at low heads and high
heads through sands, silts, clays, and broadly graded glacial soils.
In each case, a filter having a closely controlled gradation was used
downstream of the soil. Various filters of sand or gravel, or both,
were tested with each soil to determine the filter grain size distribution limits that were successful and unsuccessful in preventing
failure. Cracks or holes were formed in the soil specimens or water

pressure was used to hydraulically fracture the soil, thus simulating conditions of cracking in an earth embankment. For successful filter gradations, the test results were very dramatic and convincing. Shortly after high water pressure caused an initial surge of water through the filter, the successful filters diminished the flow to a clear trickle or drip. Unsuccessful filters allowed the flow to gradually increase, become extremely cloudy, and eventually, wash the finer portion of the soil through the filter.

The laboratory studies show that the size of the pore channels in a filter which govern it's ability to restrict the movement of soil particles is determined by the D_{15} size of the filter (13). The D_{15} of the filter at the point between successful and unsuccessful filters was determined and noted as the D_{15B} (boundary between successful and unsuccessful filters).

For fine silt and clay soils the boundary between successful and unsuccessful filters proved very close. For the broadly graded soils, the boundary is not so well defined and success or failure is determined on the basis of preventing any significant enlargement of the hole in a soil sample. In all cases, when a factor of safety is applied to the results and a filter somewhat finer than the selected failure boundary is used, no erosion takes place and the flow of water completely or almost completely stops in a very short time (a few seconds to a few minutes).

In most of the tests using low water pressure, the filter sealed the crack immediately. Examination of the specimens after the test revealed that the hole filled completely with soft soil.

The field studies included two dams where considerable cracking had been found. One dam was in Arizona, where cracks 1/8 to 1/2 in. (3 to 13 mm) wide had been found extending from the top of the dam down to a depth of about 20 ft (6 m). The other dam was in Nebraska where deep loess soils had collapsed causing cracks 2 to 6 inches (50 to 150 mm) wide for the full height of the dam (approximately 23 ft, or 7 m). Both dams have been repaired by excavating to the depth of cracking with a trenching machine and installing a graded sand-gravel filter in the trench. Small sections of these dams were diked off so that a small pond could be created on the upstream side up to the level of the embankment crest. Instruments were placed in the filter zone and the downstream cracks to monitor water pressure. The ponds were filled with water at a rate similar to a flood event in the main reservoir. Sketches of the Arizona study installation are shown in Fig. 5. The Nebraska study scheme was similar except the temporary dike was built on an upstream berm.

In the Arizona field study, the embankment soil was a sandy lean clay (CL) and the D_{85} was 0.2 to 0.7 mm. The laboratory study results indicate a D_{15B} of the filter for this type of soil to be greater than 1.5 mm. Specifications for the filter in the repair area where the field study was made called for a maximum D_{15} of 0.7 mm (a safety factor of 2). The filter selected by the contractor has a D_{15} of about 0.2 mm, which actually provides a safety factor of about 7.

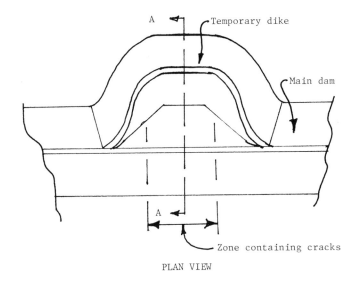

A — Temporary dike

Main dam

Zone containing cracks

PLAN VIEW

Temporary Pond

Main dam repaired
with filter

Temporary dike

SECTION A-A

FIG. 5. ARIZONA FIELD STUDY OF CRACKED DAM REPAIRED WITH FILTER

During the Arizona field study, the volume of water that entered the
filter was very low. The instruments did not detect any water in the
filter zone except adjacent to one crack where less than 0.2 ft (0.06 m)
was measured in the bottom of the filter trench after 4 weeks with full
head in the pond. Excavation of the cracks after water had been in
the temporary ponds about 30 days showed that most of the cracks sealed
at the filter face, swelled shut, or filled with loose soil that had
slaked from the sides of the crack. Some cracks excavated after the
wetting appeared to have been filled with soft soil which was then
compressed as the embankment soil swelled and closed the crack.
This field study verified the results obtained in the laboratory study.

The dam used in the Nebraska study was constructed of silt with sand
(ML), having a D_{85} of about 0.18 mm. Results of the laboratory study
showed a D_{15B} of about 1.0 mm. The repair filter in the area of the
study had a D_{15} of about 0.3 for a safety factor of about 3. During
the Nebraska study, no water flowed through the filter during filling
of the temporary pond. The water level was inadvertently allowed to
overtop the filter causing a flush of water over and through the
filter, which eroded some of the filter material from the repair
trench and the downstream crack. Subsequent excavations revealed the
cracks upstream of the filter were completely filled with fine soil.
A report of this study concludes that the filter sealed before the
overtopping occurred (1). An important conclusion is that for crack
erosion protection, the filter drainage zone should extend to a
higher elevation than the highest expected water level in the
reservoir to prevent overtopping.

The Nebraska field study also verified that a properly designed
filter having an appropriate factor of safety will seal cracks in a
embankment under field conditions.

Current Practices

Recent changes or trends in SCS practice for controlling cracking,
piping, and other related erosion problems in earth dams are as follows:

1. SCS is replacing structural cutoff walls, or anti-seep collars,
    around conduits and other structures with filter diaphragms. A
    single diaphragm is constructed around outlet conduits or other
    structures in the downstream section of the dam as shown in
    Fig. 6. The diaphragm consists of a graded sand-gravel filter
    that projects outward a minimum of 3 times its diameter (for
    pipes) except in a downward direction where differential
    settlement is not anticipated (i.e., rock). Where other
    embankment or foundation drainage systems are used, the diaphragm
    must tie into these systems to provide a continuous zone that
    intercepts all areas subject to cracking, poor compaction, or
    other anomalies.

2. Fredrickson (3,4) has developed a procedure for constructing
    small to medium size earth dams on deep loess soils that has
    proven successful in preventing severe embankment cracking. The
    procedure involves saturating deep loess foundation soils to
    induce settlement during construction of the embankment and

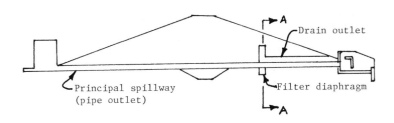

Drain outlet

Principal spillway
(pipe outlet)

Filter diaphragm

PROFILE

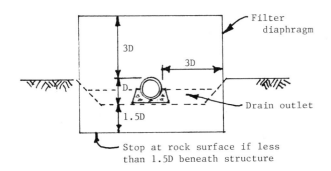

3D

3D

Filter
diaphragm

D

1.5D

Drain outlet

Stop at rock surface if less
than 1.5D beneath structure

SECTION A-A

FIG. 6.   SAND AND GRAVEL FILTER DIAPHRAGM FOR SEEPAGE AND PIPING
CONTROL AROUND STRUCTURES IN DAMS.

prevent sudden collapse at a later time when the reservoir
fills. A 2-ft (0.6 m) thick sand blanket is placed over the
undisturbed foundation area (excluding the cutoff trench area
near centerline) and covered initially with 3 ft (1 m) of
compacted embankment material. The particle size gradation of
the sand blanket is between 1 and 4 mm. Water is injected
through perforated pipes embedded in the sand blanket. With the
aid of pieometers, the water pressure in the blanket is
maintained at about 2 ft (0.6 m) of head in the beginning and up
to 5 to 10 ft (1.5 to 3 m) of head after more embankment fill
has been placed over it. Sufficient water is metered into the
system to saturate the foundation soils to the desired depth
before additional fill is placed. The embankment fill is placed
at a rapid rate and water injection is continued until the fill
placement is completed. Settlement is monitored with instruments and usually occurs as the fill is being placed, ending
soon after the dam is topped out. A typical layout of the
sand blanket is shown in Fig. 7.

The fill placement procedure reported by Fredrickson (4)
involves providing maximum flexibility of the embankment soils
so they can adjust to the foundation settlements that take place
during construction. Embankment soils are placed several
percentage points (usually 3 percent) wet of the standard
compaction optimum water content, and compaction is limited to
prevent brittleness associated with high density. Compaction is
usually accomplished with one pass of rubber-tire hauling
equipment weighing less than 70,000 lbs.(32,000 kg.). Lift
thickness is maintained between 12 and 15 in (0.3 and 0.4 m).
This procedure has produced densities between 90 and 95 percent
of standard (ASTM D 698) maximum when tested at a depth of 3 to
5 ft (1 to 1.5 m) below the working surface.

The moderate permeability of these soils allows drainage so that
the excess water added for flexibility, which may produce pore
pressure, is usually drained away from the lower reaches of the
dam by the time construction is completed. Sufficient strength
is maintained for stability.

Several successful installations using this procedure have been
made in Nebraska. It is presently used for all SCS dams in
Nebraska where deep loess foundation soils may cause excessive
rapid settlement in the foundation.

3.   SCS Engineers have become more aware of embankment cracking
      problems and the conditions that cause them. Emphasis has been
      given to finding these conditions and designing measures for
      reducing cracking potential in all dams.

      In recent years, more consideration is given to the use of
      filters as the main defense against piping that can occur due to
      cracking. SCS, as a standard practice, is not requiring filter
      installation in all high hazard dams (where loss of life in the
      event of failure is a possibility); however, such a requirement
      has been considered in light of the recent studies.

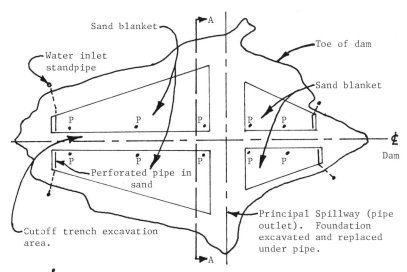

P = Piezometer installation

PLAN VIEW

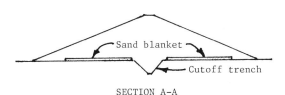

WATER INLET DETAIL

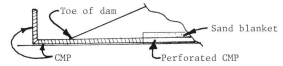

SECTION A-A

FIG. 7. SAND BLANKET DETAILS FOR SATURATING DEEP LOESS FOUNDATIONS DURING CONSTRUCTION OF DAMS

There are reasons, for and against requiring filters in all high
hazard dams. Some engineers do not agree that filters should be
required in every high hazard dam. They point out that thousands
of small dams constructed without embankment filter and drainage
zones have performed satisfactorily for many years. Some of
these may not have been economically feasible if a filter
system had been required. This applies to some high hazard as
well as low hazard dams. The flood protection provided over the
years by these dams has been very valuable, and perhaps lives
have even been saved.

Designers are always aware of the risks involved in the design
of high hazard structures and usually take a conservative
approach in treatment of soil and foundation conditions. In the
writers' opinion, the most certain defense against many common
failure causes (leakage and erosion from piping or cracking,
seepage along abutments or structure contacts, etc.) is to
install in the embankment a properly designed, continuous
filter and drainage system that extends into the foundation and
abutments. However, requiring filters for all high hazard
structures would remove the opportunity for the designer to
apply judgement on its need. Unnecessary expense and/or
elimination of projects would also frequently occur.

For the present, the use and extent of filter designs in embankment dams of all sizes and hazard classes will be a part of the
design process based on each individual site analysis. The
writers strongly recommend the use of properly designed filter
and drainage systems as the main defense against piping resulting from cracking and other related problems whenever there is
doubt regarding the safety of high hazard structures. In some
cases, the use of such systems may eliminate the need for other
expensive treatments.

4.   SCS currently is revising its criteria for filter gradation
     design to conform to the findings of recent studies on filters.
     Present criteria ($D_{15} = 5d_{85}$) are adequate and remain the same
     for sands, some silts, and mixtures of silt and sand. Changes
     will be necessary for most silts and clays and for broadly
     graded soils The details of these changes are currently under
     development and cannot be reported here.

Summary and Conclusions

Evidence indicates that nearly all dams crack to some degree, regardless of their size. This is particularly true for those constructed
on yielding foundations or where abutment configurations and embedded
structures cause uneven distribution of stress. Minor cracking may
not be detrimental for some embankment soil conditions; however,
experience has shown that dispersive clays, certain broadly graded
soils, and other erodible soils are vulnerable to concentrated
seepage, serious erosion, and possible failure.

Some SCS dams that have cracked have been repaired by trenching into the dam to facilitate installation of a filter zone. Field tests using ponded water against the repaired cracking zone have shown this method to be successful. These field studies verify the results of laboratory studies made by SCS.

Certain broadly graded soils are internally unstable because the fine portion is able to move with percolating water through the gravel size particles. These soils do not have enough sand size particles to fill the voids between the gravel-sized particles. Erosion of the fine particles by percolating water results in large sinkholes or cavities near the crest of the dam. Field evidence indicates that such erosion begins in minor cracks associated with steep abutments, structures, or other crack-causing features. The fine particles are washed through filters that are too coarse because traditional filter design criteria are not adequate for this soil.

SCS engineers have developed procedures for determining the grain size at which a filter will just begin to allow erosion in concentrated leaks and fail to seal cracks (14). Applying a factor of safety will result in a finer filter that can be used with a high level of confidence for preventing erosion.

The trend in SCS is toward using filters to ensure safety against erosion through cracks and other anomalies. For example, SCS is replacing structural anti-seep collars with filter-drainage diaphragms. Designs of large, important dams having high hazard classifications usually include drainage zones with properly designed filters as an added precaution.

APPENDIX--References

1. Dunnigan, L.P., Frederickson, R.J., Flanagan, C.P., and Wortman, W.I., "Field Test of a Chimney Filter for a Cracked Dam", USDA Soil Conservation Service, August 1983 (Unpublished).

2. Forsythe, P., "Experiences in Identification and Treatment of Dispersive Clays in Mississippi Dams", American Society for Testing and Materials, STP 623, 1977 pp. 135-155.

3. Federickson, R. J., "Foundation Treatment for Small Earth Dams on Subsiding Soils," International Association of Hydrological Sciences, Proceedings of the Anaheim Symposium, Publication No. 121, December 1976.

4. Fredrickson, R.J., "Settlement Study of Embankments and Foundations: Cone Penetrometer Tests, Laboratory Tests and Settlement Plates." May 1983 (Unpublished) USDA Soil Conservation Service.

5. Perry, J.P., "Lime Treatment of Dams Constructed with Dispersive Clay Soils", Transactions of the American Society of Agricultural Engineers, Vol. 20, No. 6, pp. 1093-1099, 1977.

6. Ryker, N.L., "Encountering Dispersive Clays on SCS Projects in Oklahoma", Proceedings, American Society for Testing and Materials Symposium on Dispersive Clays, Chicago, Illinois, 1976, pp. 370-397.

7. Sherard, J.L., "Embankment Dam Cracking", Embankment--Dam Engineering, John Wiley & Sons, Inc., New York, NY., 1972, pp. 271-354.

8. Sherard, J.L., "Study of Piping Failure and Erosion Damage from Rain in Clay Dams in Oklahoma and Mississippi", USDA, Soil Conservation Service, Washington, D.C., March 1972 (Unpublished)

9. Sherard, J.L., "Sinkholes in Dams of Coarse Broadly Graded Soils", 13th ICOLD Congress, India, Vol. II, 1979, pp. 25-35.

10 . Sherard, J.L., Discussion of "Design of Filters for Clay Cores for Dams", by P.R. Vaughan, and Soares, H.F., Journal of the Geotechnical Division, ASCE, September 1983, pp. 1195-1196.

11. Sherard, J.L., Decker, R.S., and Ryker, N.L., "Hydraulic Fracturing in Low Dams of Dispersive Clay", Proceedings, Specialty Conference on Performance of Earth and Earth-Supported Structures, ASCE, Vol. 1, Pt. 1, 1972, pp. 563-590.

12. Sherard, J.L., Dunnigan, L.P., and Decker, R.S., "Some Engineering Problems with Dispersive Clays", American Society for Testing and Materials, STP 623, 1977, pp. 3-12.

13. Sherard, J.L., Dunnigan, L.P., and Talbot, J.R., "Basic Properties of Sand and Gravel Filters", Journal of Geotechnical Engineering, ASCE, Vol. 110, No. 6, June 1984, pp. 684-700.

14. Sherard, J.L., Dunnigan, L.P, and Talbot, J.R., "Filters for Silts and Clays", Journal of Geotechnical Engineering, ASCE, Vol. 110, No. 6, June 1984, pp. 701-718.

15. Sherard, J.L., Ryker, N.L., and Decker, R.S., "Piping in Earth Dams of Dispersive Clay", Proceedings, Specialty Conference on the Performance of Earth and Earth-Supported Structures, ASCE, Vol. 1, pt. 1, 1972, pp. 589-626.

16. Sherard, J.L., Ryker, N.L., and Decker, R.S., "Hydraulic Fracturing in Low Dams of Dispersive Clay", Proceedings, Speciality Conference on the Performance of Earth and Earth-Supported Structures, ASCE, Vol. 1 pt. 1, 1972, pp. 653-689.

17. Sherard, J.L., Ryker, N.L., and Decker, R.S., Closure to "Piping in Earth Dams of Dispersive Clay", Proceedings, Speciality Conference on the Performance of Earth and Earth-Supported Structures, ASCE, Vol. 3, 1972, pp. 143-156.

18. Steele, E.F., "Character and Identification of Dispersive Clay Soils", American Society of Agricultural Engineers, paper No. 76-2022, June 1976.

19. Stevenson, J.C., Smith R., and Sterns, C.E., "Cracking of Earth Dams in Arizona", USDA, Soil Conservation Service, Report of Crack Study Team, 1978.

20. Talbot, J.R., "Cracking of Dams in Arid Regions", American Society of Agricultural Engineers, paper No. 80-2548, December 1980.

APPENDIX II - Notation

The following symbols are used in this paper:

d_{85} = particle size in base soil for which 85% by weight of soil particles are smaller;

D_{15} = particle size in filter for which 15% by weight of particles are smaller.

D_{15B} = D_{15} size of filters found by testing to be the boundary between successful and unsuccessful tests for any given soil.

ABUTMENT SEEPAGE AT SMITHVILLE DAM

by Francke C. Walberg[1], M. ASCE; Heidi L. Facklam[2], A.M. ASCE;
Karl D. Willig[3], M. ASCE; John E. Moylan[4]

ABSTRACT

    After initial filling there was concern for the stability of
Smithville Dam when higher than anticipated piezometric levels were
experienced in the left abutment foundation. Concern heightened when
review of strength data led to uncertainty about design shear
strengths for foundation shales. The left abutment overburden
consists of a thick glacial drift deposit overlying nearly horizontal
bedrock strata. During design, glacial outwash sands and gravels
were thought to be not extensive or continuous enough to permit
potential seepage paths. An investigation consisting of exploratory
drilling, sampling, testing and instrument installation was
undertaken to determine the subsurface conditions (including the
shale shear strength), to re-evaluate the stability of the dam and to
monitor the left abutment area. The investigation revealed a
continuous basal zone of coarse, glacial outwash sediments which is
transmitting high piezometric pressures through the abutment. Shear
strength of the shale was about the same as for design. A stability
analysis showed that pressure relief wells at the downstream toe and
under the downstream slope would provide a satisfactory safety
factor.

INTRODUCTION

    Smithville Lake is a U.S. Army Corps of Engineers multipurpose
project located in northwest Missouri on the Little Platte River.
The damsite is about 5 miles (8.5 km) north of Kansas City. In
addition to providing flood control benefits, lake storage is
allocated for water supply, recreation, conservation use, and
sediment reserve. The dam is a rolled earthfill embankment with a

1. Chief, Dams and Foundation Section, Kansas City District, U.S.
   Army Corps of Engineers.

2. Civil Engineer, Dams and Foundation Section, Kansas City
   District, U.S. Army Corps of Engineers.

3. Chief, Foundations and Materials Branch, Kansas City District,
   U.S. Army Corps of Engineers.

4. Chief, Geology Section, Kansas City District, U.S. Army Corps of
   Engineers.

dike section located in a low area high on the left abutment.
Foundation material consists of valley alluvium, loess and glacial
drift. When seepage and high uplift pressure were found in the left
abutment glacial drift after initial filling there was concern for
the stability of the dam. Concern heightened when review of shale
strength data led to uncertainty about shear strengths used for
design. Design strengths which were not critical for design seepage
conditions became critical with the actual seepage conditions. The
paper compares the actual performance of the dam versus the
anticipated design performance. Subsequent investigations consisting
of exploratory drilling, sampling, testing and instrument
installation are described. Results of the investigations and the
re-evaluation of embankment stability are included. Interim measures
as well as the seepage control measures which are believed necessary
for a permanent solution are presented.

BACKGROUND

     As part of the Corps' dam safety program the dam was formally
inspected during reservoir filling. In the fall of 1982 at the first
inspection after the pool had reached multipurpose level (the normal
operating pool) high piezometric response was noted in several of
the piezometers downstream of centerline. In general, when the lake
was at multipurpose pool, El. 864.2 feet (263.6 m) piezometric levels
under the downstream slope in the left abutment foundation were
somewhat above the horizontal pervious blanket. Indications from
design documents (1) were that piezometric levels for stability
studies were assumed to be at the base of the horizontal pervious
blanket downstream of centerline. Based on this, the safety factor
could be somewhat lower than the 1.6 obtained during design for the
steady seepage case. Close monitoring of piezometric levels was
continued, particularly on the left abutment (see Figure 1) where the
level was about 3 feet (.9 m) above ground surface at the toe.

     In April 1983 a site visit was made to inspect the dam with a
record high pool at El. 869.4 feet (265.2 m). A wet area had
developed at the toe on the left abutment, however, seepage was not
enough to observe flowing quantities. After analysis of piezometric
data obtained at the higher pool level, a review of the stability
analysis performed during design was begun. In August 1983 a wet
area 3000 feet (915 m) below a dike on the left abutment was
reported, and an investigation of the left abutment foundation was
initiated.

PROJECT DESCRIPTION

     Smithville dam is a 75-foot (22.9 m) high, 4000-foot (1220 m)
long rolled earthfill, zoned embankment. It has a central impervious
core with upstream and downstream random zones and an upstream berm
zone. Internal drainage consists of an inclined pervious drain and
horizontal blanket. A typical section cut along line A-A of Figure 1
is shown on Figure 2. An uncontrolled spillway is located approximately 1700 feet (518 m) beyond the end of the embankment on the
right abutment. The outlet works consists of an intake tower, gated

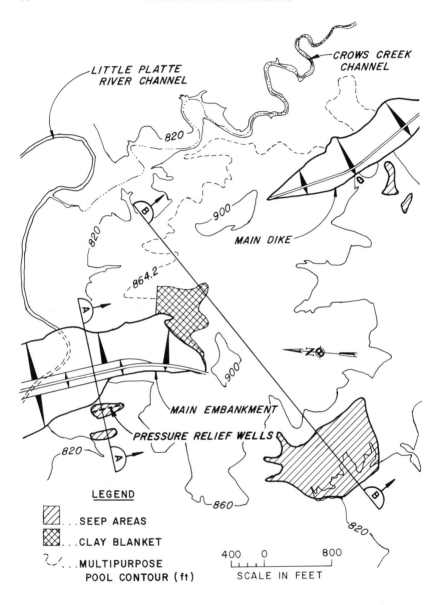

Figure I. PLAN VIEW OF SMITHVILLE DAM—LEFT ABUTMENT
(I ft = 0.305 m)

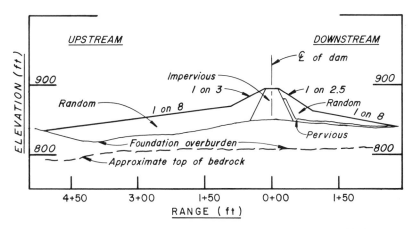

Figure 2. SMITHVILLE DAM—TYPICAL LEFT ABUTMENT SECTION
(I ft = 0.305 m)

conduit, gated water supply line, and stilling basin. An earthfill
dike on the left abutment, approximately 1/2 mile (.8 km) from the
main embankment, is 30-feet (9.1 m) high and 2400-feet (732 m) long.
The project was authorized by the Flood Control Act of 1965 with
construction starting in February 1974.   Impoundment began in October
1979 but lake filling was delayed because of real estate acquisition
problems. Multipurpose pool El. 864.2 feet (263.6 m), was first
reached in June 1982.   The record high pool, El. 869.4 feet (265.2 m),
was experienced in both April 1983 and April 1984.   Flood control
pool (full pool) is El. 876.2 feet (267.2 m).   A spillway crest pool
is El. 880.2 feet (268.5 m).   (All elevations are based upon feet/
meters above mean sea level.) Multipurpose lake storage capacity is
102,000 acre-ft (1.25 x 10^8 m^3) while flood control pool has an
additional capacity of 92,000 acre-feet (1.13 x 10^8 m^3).

GEOLOGY

     Major topographic features in the lake area are maturely to
submaturely developed valleys of the Little Platte River and its
tributaries. Drainage patterns are developed on thick glacial
deposits resulting in gently rolling topography. Bedrock exposures
are not common but can occasionally be found along the bases of
valley walls of major streams. Maximum relief in the area is about
160 feet (49 m). During the Pleistocene epoch, glaciers of the
Kansan glacial episode extended into northwest Missouri. During an
earlier Nebraskan episode, glaciers may have also advanced into the
area. Both the Nebraskan and Kansan advances were from the north-
northwest. The southern limit of glaciation is only a few miles
south of the project.

     Upland areas are deposits of glacial drift thinly mantled with
loess. In the left abutment area, drift ranges in thickness up
to 85 feet (25.9 m) and generally consists of 20 to 60 feet (6 to 18

m) of till overlying 5 to 25 feet (1.5 to 7.5 m) of coarser outwash
sediments. The till is predominantly lean clay with scattered gravel
and cobbles and occasional isolated sand lenses. Coarser outwash
sediments underlying the till are characteristically silts with some
silty clays in the upper part with a lower zone of dirty sand,
gravel, and cobbles. Alluvium occupies the valleys of the Little
Platte River and its tributaries and generally consists of lean and
fat clays overlying clayey sands and sandy clays with minor amounts
of basal gravel. Thicknesses range from 25 to 50 feet (7.5 to 15 m).

Near surface bedrock strata are of the Pennsylvanian System and
consist of alternating beds of shale and limestone. The uppermost
units are in descending order, Lane Shale, Raytown Limestone, Muncie
Creek Shale, Paola Limestone, and Chanute Shale. The top of bedrock
in the left abutment is the Raytown Limestone except where remnants
of the Lane Shale exist further into the abutment. The nearly
horizontal configuration of the left abutment bedrock surface is the
result of a pre-Pleistocene stream channel trending generally east-
west through the abutment. It is associated with the ancestral
Missouri River drainage system formed prior to the advance of
Pleistocene glaciers.

DESIGN PHILOSOPHY AND ANTICIPATED PERFORMANCE - LEFT ABUTMENT

Even though pervious outcrops were observed along the channel of
Crows Creek upstream of the dam, neither high piezometric levels nor
significant underseepage were expected. No seepage control measures,
such as impervious blanketing of outcrops, pressure relief wells or a
positive cutoff to bedrock were considered necessary because of a)
the thickness of the impervious foundation materials, b) the large
amount of fines in the sands and gravels and c) boring interpretation
which indicated pervious layers were not continuous. However, other
precautionary measures were taken. These included a piezometer
monitoring program and the following design measures. A horizontal
pervious blanket was constructed at the base of the embankment and
extended up the abutment. Its main purpose was to intercept
groundwater seepage from the foundation. Random fill consisting of
low permeability material was placed upstream of the impervious core
in the left abutment area and covered an existing draw. This draw
was believed to possibly intercept zones of sand and gravel which
might be continuous through the abutment. A 5-foot (1.5 m) thick
clay blanket completes the cover on the draw (see Figure 1). Curtain
grouting in a 400 foot (122 m) reach of the weathered rock of the
lower left abutment was conducted through the overburden. Grouting
extended through the Raytown, Muncie Creek, and Paola units into the
Chanute Shale.

The embankment was designed to provide a minimum slope stability
safety factor of 1.6 for a steady seepage case with the pool at
spillway crest. Usually a minimum safety factor of 1.5 is required
(2). Since the foundation bedrock contained weathered shale having a
stress-strain characteristic significantly different from the embankment soils, a higher safety factor was used. Analyses were also made
using the ultimate or residual strength for shale to check for a

minimum 1.0 safety factor. Bedrock was assumed to be shale even
where limestone was present. The significantly higher strengths for
the overburden or the limestone were not used since remnants of
weathered shale may be present above the limestone and the thin,
partially weathered limestone would be required to carry very large
forces. Design strength envelopes used in the stability analyses
were based on peak strengths from laboratory tests except that direct
shear tests on soft weathered shales included both peak and residual
shear strength determination.

Limit equilibrium slope stability analyses were conducted using
the Corps wedge method (2) with the base of the wedge-shaped shear
surface along the top of bedrock. The active wedge side force
inclination was assumed to act at one-half the average outer slope of
the embankment and the passive wedge side force inclination was
assumed to act horizontally. A typical left abutment embankment
section was analyzed using the phreatic surface shown on Figure 3.
The phreatic surface downstream of the inclined pervious drain was
assumed at a level well below the top of the foundation. Using peak
strengths a safety factor of 1.64 was obtained. When the residual
shear strength for shale was used the safety factor was 1.08.

PROJECT PERFORMANCE - LEFT ABUTMENT

After the seep areas at the toe of the dam and below the dike
were found, a field reconnaissance of the entire left abutment region
was made. Three general areas of seepage were identified (see Figure
1). The first seep about 3000 feet (915 m) below the dike covers an
extensive area of land adjacent to the valley alluvium of Wilkerson
Creek. The area is characterized by numerous seeps and is wet and
extremely soft. The second seep, with flows less than 1 gallon per
minute (.004 m^3/min), is located immediately downstream of
the toe of the dike. The third seep immediately downstream of the
dam embankment has no visual flow but has remained soft and spongy.
During the extended dry period local farm ponds within the seepage
area were all overflowing. Also, several mature trees located in the
large seep area have died, apparently as a result of intolerance to
the elevated zone of saturation. Prior to impoundment some of the
current seep areas contained flowing springs or became unusually wet
during periods of wet weather.

An exploratory drilling program was begun to determine in more
detail the characteristics of the left abutment foundation and to
obtain more piezometric data. Preconstruction borings indicated that
the more permeable glacial drift should be located near bedrock.
Piezometers and observation wells were installed in borings taken
into rock along line B-B on the abutment (see Figure 1), in the large
seep area, and in the embankment foundation. Borings were advanced
by continuous sampling of overburden materials with a 6-inch (15.2
cm) diameter drive barrel and cable tool drills. Representative
samples were retained for laboratory soil classification and mechanical grain size analyses. Because cobbles and coarse gravel were
encountered near bedrock, some of the borings were cored into bedrock

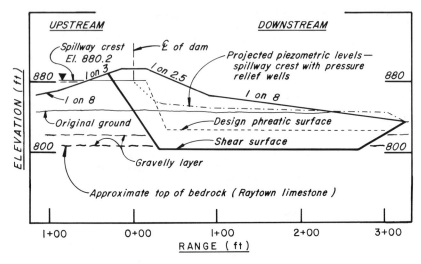

Figure 3. SMITHVILLE DAM—STABILITY SECTION A-A
(1 ft = 0.305 m)

short distances to positively identify the bedrock surface. Observation wells were used to monitor the more permeable basal layer of the glacial drift encountered above bedrock while Casagrande type piezometers were used for specific zones in the basal layer.

The basal layer of the glacial drift was found to consist of up to 40 feet (12 m) of a hetergeneous mixture of silt, and silty or clayey sands and gravels, which form a natural pervious zone. This zone contains between 10 to 30 percent fines passing the No. 200 sieve. The thickness of the zone ranges from 40 feet (12 m) under the upland portion of the abutment to less than 10 feet (3 m) towards the valleys (see profile on Figure 4). The zone terminates or becomes very thin in the valleys having been eroded away and replaced by alluvium. The material is more gravelly and cobbly where it is thickest, and becomes more sandy as it thins. While these materials were known to outcrop along Crows Creek upstream of the dam prior to construction, boring interpretations indicated they were too discontinuous and erratic to cause seepage problems after impoundment. However, now it appears the more pervious zones are indeed continuous under the entire left abutment area.

Prior to impoundment, the basal layer was a mostly saturated unconfined aquifer system whose main discharge points were the exposures along the banks of Crows Creek along with the valley alluvium and naturally occurring springs. When impoundment began, the exposed basal material along the banks of Crows Creek became submerged. The discharge area became the recharge area, initially completing saturation of the aquifer and then subjecting the confined aquifer system to a hydrostatic pressure head corresponding to the

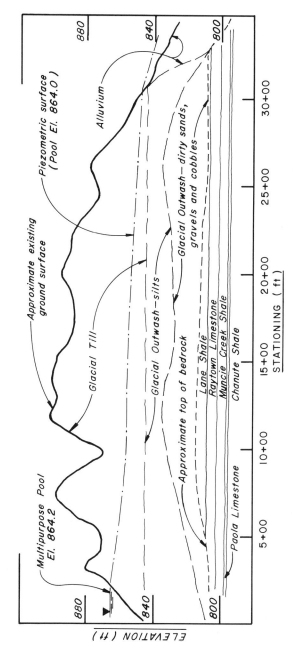

Figure 4. SMITHVILLE DAM—PROFILE B-B THROUGH LEFT ABUTMENT (1 ft = 0.305 m)

lake level. Pressure levels increased as the lake elevation rose to
multipurpose pool. Pressure levels in the basal layer near the
downstream base of the abutment increased to above ground level,
providing enough vertical gradient to force seepage through the
thinning confining layer. Seepage quantities are relatively low
since the aquifer terminates or becomes very thin. Flow from natural
springs has increased. Since the aquifer has reached a saturated
condition, changes in pool level produce rapid pressure changes in
the confined aquifer system. Data obtained during pool rises in the
spring of 1983 indicated piezometric responses up to 50% of pool
level increases with higher responses closer to the lake entrance
point.

Actual piezometric levels recorded in the embankment foundation
were some 25 feet (7.6 m) higher than those assumed for the steady
seepage case design. Using the response data these piezometric
levels were projected for spillway crest pool and used in a
preliminary slope stability re-analysis of the left abutment
embankment section. The safety factor calculated using the design
peak strengths was 1.23 and was 0.92 using the residual shear
strength for the shale.

INVESTIGATION AND REMEDIAL MEASURES - SCOPE

The low safety factors made the shear strength of shale zones
near the bedrock surface more critical. Design shale strengths, both
peak and residual, were based on laboratory tests from samples
obtained from the shale units in the right abutment, outlet works
area, and valley. The strength envelopes were controlled by lower
strengths obtained from other shale units not present in the left
abutment. Tests from left abutment shale units were limited. Good
quality shale samples from the left abutment shale units were difficult to obtain because of the thick overburden, the presence of
gravels and cobbles above rock, thinness of the shale seams, and the
weathered, jointed nature of the limestones overlying the shales.
Since the safety factor for the left abutment stability analysis had
been adequate, additional sampling and testing for the left abutment
was not warranted during design. However, the upper units in the
flat lying sedimentary rock foundation of the left abutment have been
subjected to both erosional unloading and glaciation, whereas the
same units in the outlet works area and right abutment were protected
by overlying bedrock units. Either weathering, movement of the ice
mass over the nearly horizontal left abutment bedrock surface or
valley stress relief could have caused one or more shear zones near
the bedrock surface particularly at or near the Raytown Limestone/
Muncie Creek Shale contact. The strength of a shear zone could be
less than the design strength. If movement had occurred along a
shear zone the strength could be approaching the residual condition.

Thus, in addition to higher than expected piezometric pressure in
the foundation there was an uncertainty about the available shear
strength in the foundation bedrock. The major concern was for the
stability of the dam if pool levels exceeded those already experienced and particularly as the lake level approached full flood pool.

The seep areas below the dike are not a dam safety concern, but
rather a maintenance problem. The overlying clayey materials are not
susceptible to piping and the seepage paths are very long. The
investigations and analyses conducted and remedial measures undertaken to insure the safety of the dam are discussed below.

Initially, consideration was given to increasing stability with
an emergency berm. Wet conditions at the site dictated that the berm
would be most easily constructed from rockfill obtained offsite. A
rockfill berm, besides being costly, would have made additional
exploratory drilling (necessary to determine the shale strength)
difficult or impossible. Therefore, an alternative plan was
developed in lieu of constructing on emergency berm. Included were
the following measures:

1. Immediately install slope inclinometers through the left
    abutment embankment into foundation bedrock to detect any
    zones of movement.

2. Implement a revised plan of lake releases to reduce the
    chances of subjecting the dam to high pool levels.

3. Increase the frequency of visual and instrument monitoring of
    the dam for all pool levels above multipurpose.

4. Undertake an extensive stability investigation including
    drilling, sampling, laboratory testing, and installation of
    piezometers and pressure relief test wells.

5. Develop an interim solution to improve stability at high
    pools by reducing piezometric levels by pumping from wells
    installed through the downstream embankment. Early results
    of sampling and testing would be used to assess interim
    embankment stability.

This plan minimized the risk to the dam and allowed time to determine
the foundation bedrock shear strength, assess the embankment stability, and if necessary, develop a more cost effective remedial
measures than a rockfill berm. Installation of test wells at the
embankment toe would allow evaluation of a pressure relief well
alternative.

STABILITY INVESTIGATION

    Exploratory Drilling Program

    The initial purpose of the exploratory work was to determine if a
weak zone or zones did exist. Sampling efforts were directed towards
obtaining 6-inch (15.2 cm) core samples of soft shale seams in the
Raytown Limestone, the contact of the Raytown and underlying Muncie
Creek Shale, and suspected soft seams in the upper Muncie Creek where
persistent core losses had occurred in previous borings. Detection
of soft zones, shear planes or slickensides in the remainder of the
Muncie Creek Shale or in the underlying Chanute Shale was desired.

Five-inch (12.7 cm) fixed piston Shelby tube sampling was also conducted near the base of the abutment where it was suspected the Muncie Creek might be soft and highly weathered because of the absence of the overlying Raytown Limestone in this area.

Several thin zones of core loss and "spins" (rotary shearing of the core on horizontal planes) in the Muncie Creek and Chanute Shales were found in initial borings. To determine if these resulted from the drilling process or represented natural weak zones, an overcoring technique was developed. An NX (2-1/8-inch or 5.4 cm diameter) core hole was drilled through the Chanute Shale and filled with cement grout. The grout was allowed to cure for approximately 60 hours. The NX hole was then overcored with a 6-inch core barrel. If thin weak zones were present, the rigid column of grout would provide sufficient torsional shear resistance to prevent "spins" and core losses. Core from the NX hole, the 6-inch overcore and an adjacent 6-inch core hole were carefully compared to determine if spins and core losses represented shear zones. Suspected soft zones or slickensides near the top of the Muncie Creek were not present in the 6-inch overcore. The Raytown/ Muncie Creek contact was firm and unsheared and the remaining Muncie Creek and Chanute Shales contained no presheared or soft zones.

In every core hole, however, a very soft continuous shale seam containing one or more slickensided planes was found in the lower part of the Raytown Limestone. The nearly horizontal seam, about 0.5 feet (15 cm) thick, is located approximately 1-1/2 feet (.5 m) above the Muncie Creek. Apparent continuous slickensides were observed in every sample of the shale seam recovered. Several thin shale partings are present above the seam but are not continuous.

Shear Strength Testing

Consolidated-drained (CD) direct shear tests, residual shear tests, and triaxial compression consolidated-undrained (CU) tests (with pore pressure measurement) (3) were conducted. Residual shear specimens were tested in a three-inch (7.6 cm) square direct shear box. The residual shear condition is attained at large deformations by repeatedly reversing the direction of shear on the induced shear plane (3). Triaxial compression CU tests were conducted on the slickensided surface in the Raytown shale seam to better define the effective stress envelope (especially any cohesion intercept) and to develop a total stress strength envelope. Triaxial specimens were oriented so that the slickenside was 55° to 60° from horizontal to assure strength was measured on the "slick". Specimen size was 1.4-inch (3.6 cm) diameter by 3-inch (7.6 cm) long so that 2 specimens could be trimmed from a single 6-inch (15.2 cm) core.

The first direct shear test on the Raytown seam had a higher than expected strength but later was found to have a concave, rather than a planar shaped, "slick" surface. Additional triaxial and direct shear tests were performed to increase the confidence level in the strength results. Peak strength for the foundation bedrock was conservatively selected from both triaxial and direct shear test

results from the Raytown shale seam. Approximately two-thirds of the
test values are above the selected strength. The selected strength
was not significantly lower than the design peak envelope (see Figure
5), although the residual strength was slightly lower than before,
dropping from $\emptyset_r$ = 9.0° to $\emptyset_r$ =7.4°.

STABILITY ANALYSIS CRITERIA

    Strength Approach and Required Safety Factor

    The left abutment foundation contains a weak slickensided shale
seam. The stress-strain characteristics for this seam exhibit a
significant drop in shear strength once peak strength is reached.
Because a weak seam of this type increases the possibility of
progressive failure one has three possible choices in proceeding with
the stability analysis: (a) use peak strengths in the analysis but
increase the required safety factor above the 1.5 usually required
(for the steady seepage case), (b) use peak strengths for the shale
but significantly reduce the strength in the other materials, or (c)
use the ultimate or residual strength for the shale. The second
approach was selected which allowed use of the peak strength for
shale but which required conservatism in selecting less than peak
strengths for all other materials. Strengths were selected at 0.5%
strain for the passive wedge portion of the shear surface since large
deformations are required to develop full passive resistance. The
second approach; rather than the first which increases the safety
factor, was believed to provide better insight into the degree of
conservatism introduced into the analysis. The last options use of
residual strength was believed to be overly conservative.

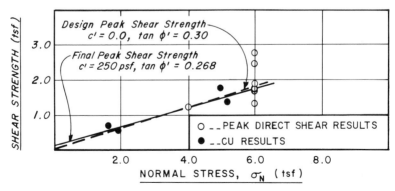

Figure 5. PLOT OF BEDROCK STRENGTH—RAYTOWN LIMESTONE
SHALE SEAM (I tsf = 95.76 kN/m²)

The Corps has considerable design experience using its wedge
method. It is a force equilibrium procedure in that force
equilibrium is satisfied but moment equilibrium is not. As a result
the calculated safety factor is sensitive to the inclination of
interslice side forces.(4) In our experience when the active and
passive wedge side force inclinations are assumed to be horizontal
the safety factor has typically been about 0.2 less than a
"mechanically correct" method, such as Spencer's procedure, that
satisfies all conditions of equilibrium. Rather than keep the
required safety factor at 1.5 and assuming a steeper side force
inclination (which would have an undetermined effect on the safety
factor) it was proposed to use horizontal active and passive side
force inclinations and reduce the required safety factor from 1.5 to
1.3. Use of conservative projections for piezometric surfaces for
the analysis (discussed in more detail below) and the fact that
foundation conditions, including shale strengths, were now quite well
determined both support the reduction in required safety factor. As
an added precaution the analysis was checked using Spencer's
procedure. Spencer's procedure, besides satisfying both force and
moment equilibrium, can be used to analyze non-circular wedge-shaped
shear surfaces. The calculated safety factor is insensitive to side
force inclination. The method has been shown to yield accurate
safety factors (4).

Analyses were conducted for the steady seepage case with the pool
at spillway crest. Since the investigation revealed the strength of
the shale seam was about the same as had been used for design it was
necessary to conduct analyses for an interim solution with pumped
wells through the embankment slope as well as for a permanent
solution with pressure relief wells at the embankment toe.

Shear Strength Selection

For the interim and final analyses the peak strength was used on
the shale seam, but on the remainder of the shear surface strengths
less than the peak value were used. Because relatively large
displacements would be required to develop full passive resistance,
strengths in the passive wedge (foundation overburden) were used
which correspond to 0.5% strain, a strain level more compatible with
development of peak strength in the shale. Although tests from
samples from the completed embankment indicate higher strengths than
those used for the design analysis, the lower design embankment
strengths were used. As a result of extensive exploratory borings
through the foundation overburden, it was concluded that a coarse-
grained pervious layer was consistently present immediately above
rock. Since design tests on overburden samples had been on the
weaker lean and fat clays a somewhat higher (although not the total
available) strength was assigned to the pervious layer in the active
wedge portion of the failure surface. In summary, for both the
embankment and the overburden in the active wedge the full available
peak strength was not used in the analysis. Additionally, the
Raytown Limestone above the shale seam in the active wedge was
assumed to have a vertical joint and was given no shear strength.

See Table 1 for a summary of the design strengths used for the analysis.

Projected Piezometric Levels Determination

For the interim solution, piezometric levels at high pools would be reduced by pumping from 6-inch (15.2 cm) diameter wells installed through the downstream slope. Three wells were installed, developed and then pump tested at rates up to 25 gallons per minute (0.1 m^3/min). Results of the pump tests indicate significant lateral variation in permeability and/or thickness of the basal pervious material. Computed values of transmissivity vary from 300 to 4300 gallons per day per foot (3.7 to 53.4 m^3/day/m). The transmissivity of the pervious appears to be greatest further into the abutment but is quite small only 400 feet (120 m) away (towards the valley). To determine drawdown for use in the stability analysis, a long term pump test (27 hours with continuous pumping from all 3 wells) was conducted while monitoring the piezometers in the pervious overburden. Distance-drawdown plots were constructed from the data and used to contour cones of depression. These contours were subtracted from contours of the piezometric surface projected to the spillway crest pool to provide the piezometric surface for use in the interim stability analysis. This surface is conservative since drawdowns with the spillway crest pool should be greater than the drawdowns obtained from the pump test with the pool close to multipurpose level.

Later, four 4-inch (10.2 cm) diameter pressure relief test wells were installed at the toe of the embankment (see Figure 1) to determine if pressure relief wells would provide a satisfactory permanent solution. Following development, flow from the pressure relief test wells stabilized at rates ranging from 2 to 6 gallons per minute (.008 to .024 m^3/min) with total flow approximately 13 gallons per minute (.052 m^3/min). Piezometers in the pervious overburden were monitored to determine their response to the wells. Contours developed from drawdown data with the pool at El. 868 feet (263.9 m) were subtracted from projected piezometric contours for a

Table I. SHEAR STRENGTH PARAMETERS

ANALYSIS	DESIGN		FINAL	
PEAK STRENGTH	c'(psf)	tan ϕ'	c' (psf)	tan ϕ'
EMBANKMENT	0	0.45	0	0.45
OVERBURDEN-FINE GRAINED	0	0.45	0	0.45
OVERBURDEN-GRAVELLY	0	0.45	0	0.577
JOINT IN RAYTOWN LIMESTONE	~	~	0	0.00
BEDROCK STRENGTH	0	0.30	250	0.268
PASSIVE WEDGE- 0.5% STRAIN	~	~	40	0.13

NOTE: I psf = 47.9 N/m^2

pool at spillway crest even though drawdowns would be higher for the pool at spillway crest. The resulting quite conservative piezometric surface was used in the final stability analysis.

Stability Analysis Results

The stability analyses were conducted with pool levels at multipurpose pool and at spillway crest pool. At spillway crest pool, analyses were conducted without any wells, with pumped wells (interim solution) and with pressure relief wells (final solution). The analyses were conducted for several sections on the left abutment. A search was made for each case to determine the most critical shear surface and the minimum safety factor. Results of these stability studies for the most critical section are summarized below:

	Safety Factor	
	Wedge Procedure	Spencer's Procedure
Multipurpose Pool	1.40	1.66
Spillway Crest Pool without wells	1.20	1.41
Spillway Crest Pool with pumped wells	1.32	1.52
Spillway Crest Pool with pressure relief test wells	1.30	1.53

The safety factor for spillway crest pool without pumping is below the required 1.3, thus the interim solution was implemented. From a plot of safety factor versus pool level a determination was made that pumping would begin whenever a pool of El. 872 feet (266 m) or higher is forecast to assure a safety factor of greater than 1.3.

While the interim solution provides an adequate safety factor, it is not desirable as a permanent solution since it requires a specific human response for the life of the project whenever high pools are experienced. There are concerns that at some time in the future, when current personnel are no longer around, that the necessity of maintaining and of pumping the test wells each time the pool approaches El. 872 feet (266 m) may be forgotten. The solution is also dependent on complete mechanical and electrical reliability of pumps and generators; any significant downtime during high pools would be potentially dangerous.

The permanent solution was developed based upon the results of the four "test" wells at the embankment toe. They provide the minimum required safety factor even though the piezometric levels used in the analysis are conservative. However, the wells will lose efficiency with time. Added safety will be provided by installing more pressure relief wells at the toe and by lowering the outfall of the wells with the installation of a buried collector pipe. Additionally since the seepage gradient is almost parallel to the dam axis flowing wells will be installed through the embankment with the "outfall" in the horizontal pervious drain. The "outfall" would be provided by a well screen inserted in the riser section at the pervious drain. These wells will be located where the projected gradient is still above the base of the pervious drain after completion of the pressure relief wells at the toe.

CONCLUSIONS

The following conclusions are made.

1. Seepage is a result of influence of the impounded water of Smithville Lake on a confined aquifer system composed of coarser pervious materials with overlying impermeable fine-grained silty clays and clayey silts. The natural pervious material is very heterogeneous.

2. The confined aquifer system is in equilibrium for a multipurpose pool condition. Pressures in the system respond rapidly with rises in pool levels.

3. The embankment on the left abutment as constructed has an inadequate safety factor with respect to slope stability for a spillway crest pool, steady seepage case, because of higher than anticipated piezometric levels in the foundation.

4. An interim solution of pumped test wells through the downstream slope provides an adequate safety factor until a permanent solution can be implemented. Pumping would commence whenever the pool is predicted to rise to elevation 872 or higher.

5. A permanent solution of pressure relief wells at the toe area will provide an adequate safety factor for spillway crest pool, steady seepage conditions.

References

1. U. S. Army Corps of Engineers, Smithville Lake "Soil Data and Embankment Design" Design Memorandum No. 10, October 1970, U.S. Army Engineer District, CE, Kansas City, Mo.

2. U.S. Army Corps of Engineers, "Stability of Earth and Rockfill Dams," Engineer Manual EM 1110-2-1902, 1 APR 1970, U.S. Government Printing Office, Washington, D.C.

3. U.S. Army Corps of Engineers, "Laboratory Soils Testing" Engineer Manual EM 1110-2-1906, 30 November 1970, U.S. Government Printing Office, Washington, D.C.

4. Wright, S. G., "A Study of Slope Stability and the Undrained Shear Strength of Clay Shales," Ph.D. Thesis, University of California, Berkeley, 1969.

SEEPAGE CONTROL AT COCHITI DAM, NEW MEXICO

By Dwayne E. Lillard, P.E.[1]

ABSTRACT: Significant design data, design assumptions and computations, and performance of the seepage control features provided at Cochiti Dam are summarized in this paper. Cochiti Dam, located in north central New Mexico is approximately 251 feet (76.6 m) high at the maximum section and 28,600 feet (8.72 km) long. The dam is a rolled and zoned earthfill embankment. Volume of underseepage under permanent pool levels must be limited to a practical minimum to assure continuity of permanent pool storage. For flood control pools above permanent pool levels, the volume of underseepage is not critical, but seepage control to insure safety of the structure is paramount. Evaluations of in situ permeabilities of the granular foundation materials were complicated by variations in thickness of overburden above bedrock, the complex and irregular stratifications of granular foundation materials, and the major thicknesses of foundation overburden above existing ground water table. Based on the results of field and laboratory tests, a design permeability curve as related to D_{20} soil size was developed for seepage analyses. Based on seepage measurements during storage of the 1979 record pool, the measured foundation seepage quantities are less than the quantities of seepage estimated in design.

General

Purpose. The purpose of the paper is to provide a summary of significant design data, design assumptions and computations, and performance of the seepage control features provided at Cochiti Dam as monitored by instrumentation during initial reservoir filling and storage of the 1979 record pool. This paper was prepared from data contained in the various design memoranda for Cochiti Dam, as-built drawings, periodic inspection reports, and reports prepared for and by the Albuquerque District Office, U.S. Army Corps of Engineers.

Project Description. Cochiti Dam is located in north-central New Mexico, in Sandoval County. The dam is about 50 miles (80.5 km) upstream or north of the metropolitan area of Albuquerque, New Mexico, and approximately 25 miles (40.25 km) south of Santa Fe, New Mexico. Cochiti Dam was constructed for flood control and carryover pool, sediment detention, irrigation, fish and wildlife, and recreation purposes.

---

[1] Civil Engineer, U.S. Army Corps of Engineers, Albuquerque District, Albuquerque, New Mexico.

The project consists of a rolled earthfill embankment of approximately 64,600,000 cubic yards (49,419,000 m^3) of compacted fill with a crest length of about 28,600 feet (8.72 km) and a crest width of 30 feet (9.15 m). Design top of the earth dam is at elevation 5479, approximately 251 feet (76.6 m) above the Rio Grande streambed and about 300 feet (91.5 m) above the bottom of the cutoff trench across the Rio Grande. The dam extends generally in an east-west line across the Rio Grande and Canada de Cochiti to a point about 2 miles (3.22 km) east of the Rio Grande and then southward across the Santa Fe River. Height of dam across Canada de Cochiti is about 160 feet (48.8 m) and about 170 feet (51.8 m) across Santa Fe River. Height of dam in the low dam area, station 144+00A to 216+00A, is generally less than 50 feet (15.2 m). General plan of the embankment is shown on Figure 1. An uncontrolled concrete spillway with a 460-foot-long (140.3 m) ogee weir and a 160-foot (48.8 m) notch in the center, about 10 feet (3.05 m) deep, is located at the southernmost end of the embankment on the south side of the Santa Fe River. Operational releases for flood control and irrigation are made through the outlet works, a three-barrel gated conduit ending in a two-level stilling basin. The outlet works is located in the left abutment of the Rio Grande. No outlet was provided on the Santa Fe River. Dead storage in the Santa Fe River and the Canada de Cochiti arms of the lake was eliminated by means of a conveyance channel, closely paralleling the dam, connecting the Santa Fe River and the Rio Grande channels. Flood control capacity is 602,000 acre-feet (740.46 Mm^3) at elevation 5460.5. Permanent pool capacity is 50,000 acre-feet (61.5 Mm^3) at elevation 5322.

Embankment Zoning. Zoning in general consists of a central impervious core flanked by shells which increase in permeability toward the exterior slopes. Upstream zoning consists of random granular fill adjacent to the impervious zone with a natural pervious fill shell toward the upstream slope. Downstream zoning generally consists of an inclined and horizontal drain of processed drain material between natural pervious fill, a random fill zone and an outer shell of random granular fill. Portions of the dam of lesser height from station 2+70A to 14+00A, station 139+00A to 216+00A and station 260+32.63A to end of dam station 288+15A, consists of a central impervious zone flanked by natural pervious fill shells. Where upstream impervious blankets are required for seepage control, the central impervious zone is connected to the upstream blanket by an impervious zone constructed on foundation grade. Typical sections of the embankment are shown on Figures 2 and 3.

Seepage and Seepage Control

General. Control of underseepage at the Cochiti site is a major problem. Depths of foundation overburden over bedrock varies from zero on the left abutment of the Rio Grande and in the Santa Fe River to more than 200 feet in other areas along the alignment. Sandstone bedrock outcrops on the left abutment of the Rio Grande and basalt is exposed in the Santa Fe River canyon. The major length of dam is founded on overburden. The foundation overburden, in general, consists of a variable thickness surface mantle of finer grained soil underlain by granular materials. The granular overburden in

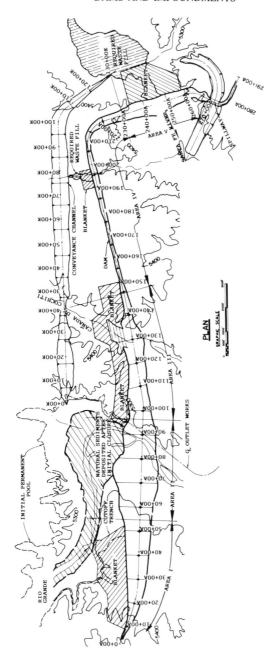

Fig. 1. - General Plan of Cochiti Dam (1ft=0.305m) (From Ref. 2)

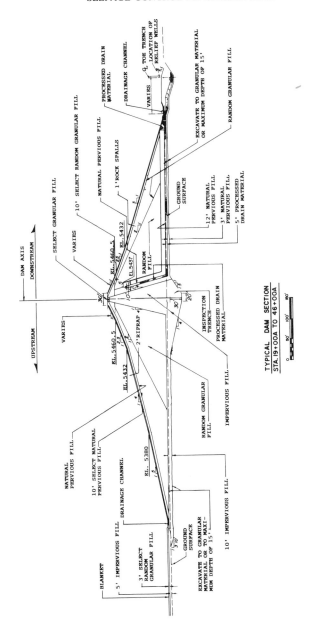

Fig. 2. - Typical Dam Section (1ft=0.305) (From Ref. 1)

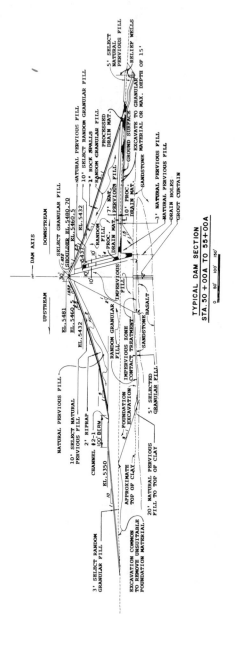

Fig. 3. - Typical Dam Section (1ft=0.305m) (From Ref. 1)

embankment foundation areas ranges from alluvial terrace sands and gravels well above present river level to more recent alluvium in the present river channel. In general, the granular foundation materials consist of relatively clean sands and gravels with random and minor lenses, pods and thin layers of fine grained material.

The project has a permanent pool of 50,000 acre-feet (61.5 Mm^3) for conservation and development of fish and wildlife resources and for recreation. Initial permanent pool elevation 5322 is contained in the Rio Grande arm of the reservoir. The permanent pool, after 50 years sediment depletion, will be at elevation 5370 which would put about 15 feet (4.58 m) of water through the conveyance channel into the Santa Fe River arm of the reservoir. Volume of underseepage under permanent pool levels must be limited to a practical minimum to assure continuity of permanent pool storage. For flood control pools above permanent pool levels, including maximum carryover pool elevation 5432, the volume of underseepage is not critical, but seepage control to insure safety of the structure is paramount. Seepage studies and estimates of underseepage volumes are based on pool levels at elevation 5322, 5370, 5432, and 5474 (maximum pool).

Permeabilities of Foundation Materials. Evaluations of in situ permeabilities of the generally granular foundation overburden at the site were complicated by:

a. variations in thicknesses of overburden above bedrock;

b. the classification range of overburden materials varying from fine to coarse grained sand and gravel to low to high plasticity clay.

c. the complex and irregular stratification of granular foundation materials; and

d. the major thickness of foundation overburden above existing ground water table.

Field and laboratory testing to determine permeability characteristics of the foundation overburden icluded 3 field pumping-out tests (PO-1, PO-2, and PO-3) in wells below water table, 3 field pumping-in tests (PI-1, PI-2, and PI-3) in wells above water table, test pumping the operations area well, numerous field pumping-in tests (constant and falling head tests) on well point-type piezometers, and laboratory permeability tests on 27 composite samples of granular foundation and borrow materials. Details and results of the field and laboratory testing are discussed in detail in references 2, 3, and 8. Locations of pumping-out tests PO-1 and PO-2 are shown in plan and profile on figures 4 through 6.

The average coefficient of permeability, K_H, of the aquifer at test well PO-1 was computed for the formulas for both artesian and gravity flow as given below.

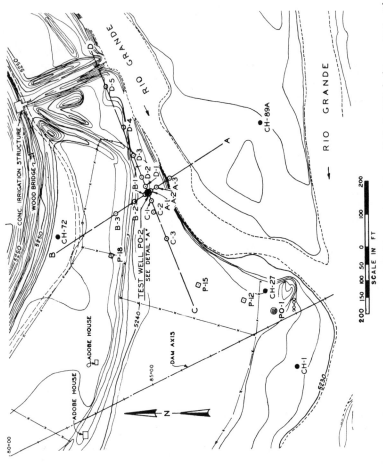

Fig. 4. – Plan of Test Well PO-2 & Piezometer Lines (1ft=0.305m) (From Ref. 8)

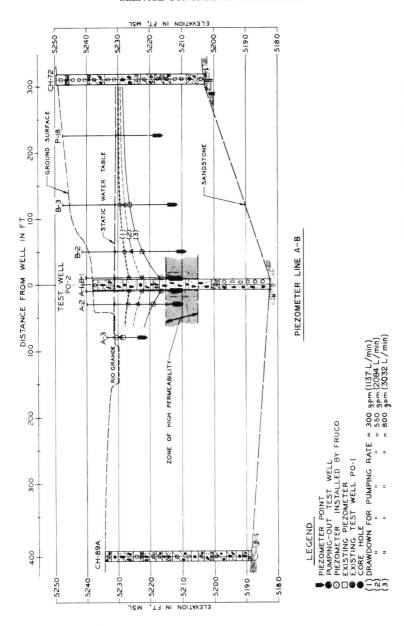

Fig. 5. - Profile of Test Well PO-2 and Piezometers (1 ft=0.305m) (From Ref. 8)

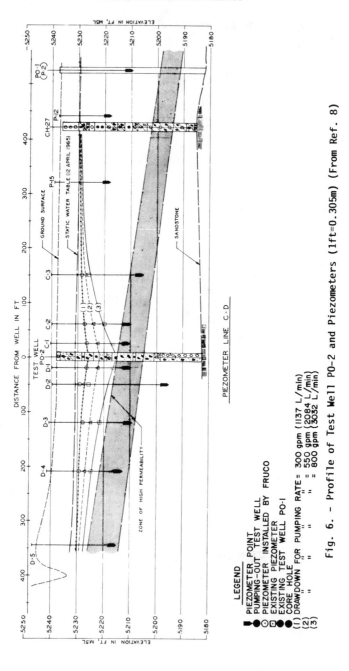

Fig. 6. - Profile of Test Well PO-2 and Piezometers (1ft=0.305m) (From Ref. 8)

Fully Penetrating Artesian Well

$$K = \frac{Q_w \; \ln \; (R/r_w)}{2 \pi D \; (H-h_w)} \quad \ldots \ldots \ldots \ldots (1)$$

Fully Penetrating Gravity Well

$$K = \frac{Q_w \; \ln \; (R/r_w)}{\pi \; (H^2-h_w^2)} \quad \ldots \ldots \ldots \ldots (2)$$

Where

Q_w = Total well flow

H = Height of initial water table

h_w= Head at the test well

$H-h_w$ = Drawdown in test well

R = Radius of influence

r_w= Radius of test well

D = Thickness of pervious stratum

The average coefficient of permeability, K_H, of the aquifer at test well PO-2 was computed from the formula for steady artesian flow, given below and the flow net, based on piezometer data, shown on figure 7.

$$K_H = \frac{Q_w}{(H - h_w) \; D} \quad \frac{N_e}{N_f} \quad \ldots \ldots (3)$$

Where Q_w, H, h_w, $H-h_w$, and D are defined above and
N_e = Number of equipotential drops
N_f = Number of flow channels

The average coefficient of permeability, K_H, of the aquifer at test well PO-3 was computed from formula (1) above for steady artesian flow.

   Results of field and laboratory tests were indicative of the range in permeabilities of the granular foundation materials at the site.   In order to utilize grain size data from subsurface investigations as a basis for estimating foundation permeabilities in specific areas along the length of dam, an attempt was made to relate D_{10} and D_{20} soil size to probable permeability. Preliminary studies indicated that use of D_{20} size gave a better fit and efforts to relate D_{10} size were ceased. The design permeability curve for Cochiti foundation overburden materials, as related to D_{20} size, is shown on

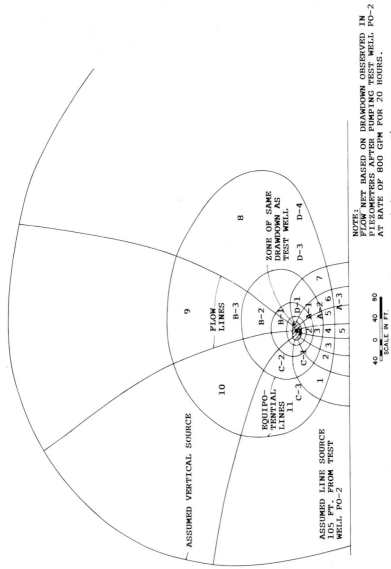

NOTE:
FLOW NET BASED ON DRAWDOWN OBSERVED IN PIEZOMETERS AFTER PUMPING TEST WELL PO-2 AT RATE OF 800 GPM FOR 20 HOURS.

Fig. 7. - Flow Net (1ft=0.305m) (From Ref. 8)

Figure 8. The curve presented was selected to encompass a large percentage of the laboratory data and correlate with results from pumping in and pumping out tests. The curve is considered to be a reasonable approximation to a maximum permeability of about 1 foot per minute (0.508 cm/sec). Sample data and data from pumping-out and pumping-in tests indicate that there may be areas or zones of greater permeability in the granular overburden. Complex and irregular stratification in the mass, however, will tend to limit the extent of such areas. It is improbable that continuous layers of highly permeable materials extend through the foundation area.

Using D_{20} sizes from laboratory gradation tests on overburden samples from an investigation hole and D_{20}-permeability curve, a weighted average permeability was determined for each sample hole. This data, supplemented or modified by field data from pumping-in and pumping-out tests, if in the area, was used to estimate an average permeability value for the foundation overburden in each area or sub-area of dam length for seepage studies. Maximum permeabilities computed for pumping-out tests ranged from approximately 6.5 feet per minute (3.30 cm/sec) to 0.29 feet per minute (0.147 cm/sec). Maximum permeabilities computed from pumping-in tests into overburden above water table ranged from 0.03 feet per minute (0.015 cm/sec) to 0.009 feet per minute (0.005 cm/sec). Weighted average permeabilities for the depths tested ranged from 0.01 feet per minute (0.005 cm/sec) to 0.006 feet per minute (0.003 cm/sec). Permeabilities calculated from constant and variable head tests on well point piezometers varied from 0 to 0.37 feet per minute (0.188 cm/sec).

Permeability tests of foundation bedrock were made by test-pumping core holes CH-32, CH-48, and CH-66 drilled into basalt in the Santa Fe River area, test-pumping core holes CH-135, CH-147, and CH-154 drilled into sandstone or basalt in the area northwest of the Rio Grande, and pressure testing and bailing other core holes. The average coefficient of permeability for CH-135, CH-147, and CH-154 was computed using formulas (1) and (2) above for a fully penetrating artesian well and a fully penetrating gravity well, respectively.

Studies of seepage losses through bedrock used permeabilities of 350×10^{-4} feet per minute (0.018 cm/sec) for basalt and 6×10^{-4} feet per minute (3.05×10^{-4} cm/sec) for sandstone. The above values are average and particular zones of basalt or faulted zones of sandstone may be more open.

Estimated Seepage Quantities

The length of dam was divided into five areas for seepage estimates. The areas are shown in plan on figure 1. Foundation overburden and bedrock conditions are unique in each area. The seepage quantities and method of analyses for each of the five areas are discussed in detail in reference 2. Typical sections, flow net analysis and method of analyses for area I are presented on figures 9 through 11. Seepage control features, method of analyses and range of estimated foundation seepage quantities for areas I, II, III, IV, and V are summarized on figure 12. Ranges of total estimated foundation seepage quantities for length of dam under different pool elevations are as follows:

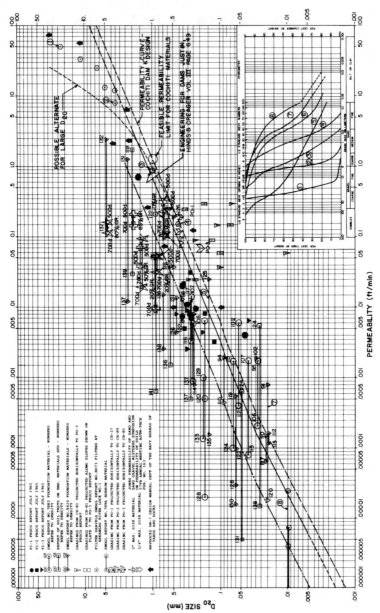

Fig. 8. – Design Permeability Curve (1ft/min.=0.508 cm/sec.) (From Ref. 2)

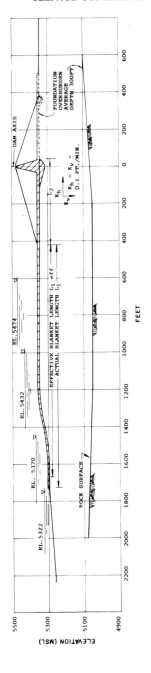

Fig. 9. - Typical Section - Area I (1ft=0.305m) (From Ref. 2)

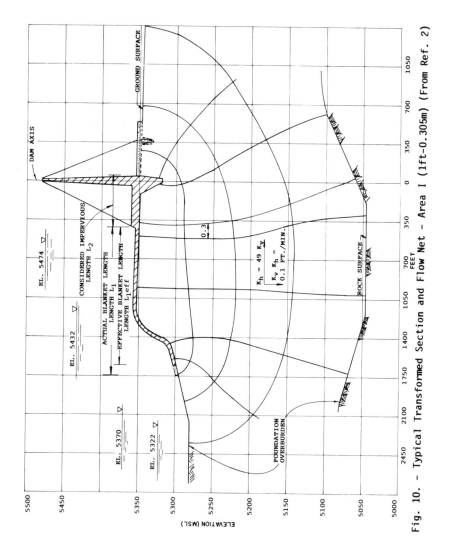

Fig. 10. – Typical Transformed Section and Flow Net – Area I (1ft–0.305m) (From Ref. 2)

METHOD OF ANALYSIS   AREA I

1. THE ANALYSIS PRESENTED IS A COMPOSITE SOLUTION USING "BENNETT'S FORMULA" AND FLOW NETS. THE VALUES PRESENTED COMPRISE THE ANTICIPATED RANGE OF SEEPAGE FOR THIS AREA. BENNETT'S FORMULA WAS USED TO COMPUTE THE MAXIMUM EFFECTIVE LENGTH OF THE BLANKET FOR BOTH ANALYSIS. THE FLOW NETS WERE DRAWN FOR AN ASSUMED STRATIFICATION RATIO OF $K_h=49 K_v$. ASSUMED PERMEABILITY FOR THIS AREA OF 0.1 FT/MIN. IS AN AVERAGE VALUE OBTAINED FROM THE D_{20} CHART USING GRADATION CURVES OF COREHOLES IN THE AREA AND FROM THE RESULTS OF PUMPING OUT TEST PO-3.

2. SEEPAGE IN THE AREA FROM STA. 40+00A TO STA. 53+00A IS CALCULATED ONLY FOR SEEPAGE THAT WOULD PENETRATE THE BLANKET UNDER A SUSTAINED POOL ELEVATION. ACCESS TO UNDER SEEPAGE IN THIS AREA HAS BEEN PREVENTED BY A CUTOFF TO ROCK (SEE PLATE 2).

3. THE ADDITIONAL FLOW RESULTING FROM INSTALLATION OF THE PRESSURE RELIEF SYSTEM IS NOT INCLUDED IN THIS ANALYSIS.

4. THIS AREA IS ABOVE THE INITIAL POOL LEVEL 5322 THEREFORE SEEPAGE WAS NOT ESTIMATED FOR THIS WATER LEVEL. SEEPAGE QUANTITIES FOR THIS POOL LEVEL SHOULD BE MINOR BECAUSE OF THE LONG SEEPAGE LENGTH.

TYPICAL CALCULATION:
BENNETT'S FORMULA

$$L_{1eff} = \frac{\tanh(aL.)}{a} \qquad a = \sqrt{\frac{K_b}{K_f\, z_f\, z_b}}$$

$$a = \sqrt{\frac{1 \times 10^{-5} \text{ FT/MIN.}}{(0.1 \text{ FT/MIN})(300 \text{ FT.})(5 \text{ FT.})}} = 6.66 \times 10 = 2.582 \times 10^{-4}$$

$$L_{1eff} = \frac{\tanh(2.582 \times 10^{-4})(1310)}{2.582 \times 10^{-4}} = 1262 \text{ FT.}$$

POOL ELEV. 5474

$$q = \frac{(K_f)(H)(z_f)}{L_1 + L_2} = \frac{(0.1)(126)(300)}{1262 + 480} = 2.17 \text{ CFM/FT.}$$

FLOW NET - STRATIFICATION RATIO 40 TO 1

$$K_h = 40\, K_v \qquad \bar{K} = (0.1)(0.1/40) = \frac{0.1}{7} \qquad q = (H)(\bar{K})\frac{K_f}{K_e}$$

POOL ELEV. 5474 $q = 126(0.1/7)\left(\frac{4}{7.3}\right) = 0.90 \text{ CFM/FT.}$

SEEPAGE IN AREA STA. 40+00A TO STA. 53+00A

$$Q = K\, i\, A$$

POOL ELEV. 5474 $Q = (1\times10^{-9})\left(\frac{150}{5}\right)(88680 \text{ FT}^2) = 26.6 \text{ CPM}$

K_b = VERTICAL COEFF PERMEABILITY OF BLANKET = 1×10^{-5} FT./MIN.

K_f = HORIZONTAL COEFF OF FOUNDATION = 0.1 FT./MIN.

z_b = BLANKET THICKNESS = 5 FT.

z_f = FOUNDATION THICKNESS = 300 FT.

L_1 = ACTUAL UPSTREAM BLANKET

L_{1eff} = EFFECTIVE UPSTREAM BLANKET

L_2 = BLANKET AND CORE UNDER DAM CONSIDERED IMPERVIOUS

q = SEEPAGE PER UNIT LENGTH OF DAM

H = HEAD, POOL LEVEL MINUS TAIL WATER

k_f = NUMBER OF FLOW CHANNELS

k_e = NUMBER OF EQUIPOTENTIAL DROPS

K = VERTICAL PERMEABILITY OF BLANKET

L = GRADIENT = HEAD/THICKNESS OF BLANKET

A = AREA EXPOSED TO POOL LEVEL

Fig. 11. - Seepage Analysis for Area I (1ft=0.305m; 1 cu ft/min.=0.028m³/min.) (From Ref. 2)

LOCATION	SEEPAGE REDUCTION METHOD	METHOD OF ANALYSIS	RANGE OF SEEPAGE ANTICIPATED							
			POOL EL. 5474		POOL EL. 5432		POOL EL. 5370		POOL EL. 5322	
			FLOW CFS	FLOW CFS	FLOW CFS	FLOW CFS	FLOW CFS	FLOW CFS	FLOW CFS	FLOW CFS
AREA I STA.0+00A TO STA.53+00A	UPSTREAM BLANKETS	BENNETT'S FORMULA, FLOW NETS, HOMOGENOUS FOUNDATION AND STRATIFIED FOUNDATION $K_H = 49 \ K_V$	100	43	57	25	11	4	MINOR	MINOR
AREA II STA. 53+00A TO STA.95+00A	CUTOFF TO SANDSTONE OR BASALT, TRIPLE LINE GROUT CURTAIN IN BASALT SINGLE LINE IN SANDSTONE	DARCY'S EQUATION APPLIED TO FLOW THRU ROCK MASS	29	9	23	7	15	4	8	2
AREA III STA.95+00A TO STA.145+00A	UPSTREAM BLANKETS AND INSPECTION TRENCH ON CENTERLINE	FLOW NETS, HOMOGENOUS FOUNDATION AND STRATIFIED FOUNDATION $K_H = 49 \ K_V$	68	38	40	25	7	4	MINOR	
AREA IV STA.145+00A TO STA.230+00A	SHORT UPSTREAM BLANKET TO TIE TO NATURAL BLANKET	FLOW NETS, BENNETT'S FORMULA	118	118	35	35	MINOR		—	—
AREA V STA.230+00A TO STA.266+00A	UPSTREAM BLANKETS AND CUTOFF TO BASALT FOR MAJOR PORTION	FLOW NETS FOR FLOW THRU BASALT LAYER	16	16	10	10	3	3		
TOTAL FLOW CFS			331	224	165	102	36	15	8	2
			EL. 5474		EL. 5432		EL. 5370		EL. 5322	

Fig. 12. – Seepage Summary (1 cu ft/sec.=0.028m³/sec.) (From Ref. 2)

	Range of Total Estimated Foundation Seepage Quantities in cubic feet per second (m^3/sec)
Pool Elevation	
5322 (Initital Permanent Pool)	2(0.056) to 8(0.224)
5370 (Ultimate Permanent Pool)	15(0.42) to 36(1.01)
5432 (Carryover Pool)	102(2.86) to 165(4.62)
5474 (Maximum Pool)	224(6.27) to 331(9.27)

Ranges of total estimated seepage quantities are the results of alternate assumptions and methods of analyses in the separate areas. Average permeabilities of foundation overburden and bedrock were based on field tests and on the D_{20} - permeability curve shown on figure 8.

Estimates of foundation seepage for the length of dam are considered to be conservatively high approximations of possible seepage losses. Estimates of average permeabilities for the complex and irregularly stratified granular foundation overburden and for the sandstone and basalt bedrock are considered reasonable for the material. There is no doubt that there are zones or layers of both higher and lower permeability in the granular overburden. The same is true in sandstone and basalt bedrock. Permeabilities of scoriaceous and more fractured zones of basalt and of sandstone adjacent to fault zones may well be higher than the average values assumed in the seepage studies.

Seepage Control Features

Embankment and foundation seepage control features are shown in typical dam sections on Figures 2 and 3. The basic seepage control features provided for Cochiti Dam are briefly discussed in the following paragraphs and discussed in detail in references 2 and 4.

Impervious Core Zone. The dam has a continuous, centrally located, impervious fill core to limit seepage through the dam. The ratio of impervious zone thickness to reservoir head is approximately 0.5 or larger in all areas. Estimated seepage through the core for the length of dam varied from 1 cu ft/min (0.028 m^3/min) at pool elevation 5322 to 31 cu ft/min (0.868 m^3/min) at pool elevation 5474.

Internal Drains. Internal drains, consisting of an inclined drain immediately downstream of the impervious core zone and a horizontal drain at or about foundation grade extending from the downstream toe of the core zone to an exit face at the downstream toe of dam, are provided from station 14+00A to 133+00A and from station 216+00A to 260+32.63A. The internal drains consist of a zone of processed drain material sandwiched between zones of natural pervious fill. Drainage blankets were not required in the lower dam areas, from the right abutment to station 14+00A, from station 133+00A to 216+00A, and from station 260+32.63A to the south end of the dam, where all downstream shell material is natural pervious fill.

The inclined drain will intercept through seepage, prevent saturation of the downstream embankment zone, and provide crack protection for the central impervious core zone. The horizontal drain provides an outlet for through seepage intercepted by the inclined drain, for foundation seepage and for seepage flow intercepted by the drain holes into basalt. Permeability and seepage studies show that the horizontal drain will carry the anticipated seepage quantities and prevent saturation of higher shell materials.

Upstream Impervious Fill Blankets. Upstream impervious fill blankets, connected to the central core zone, are provided from station 6+50A to the cutoff trench extending upstream from station 53+00A (0+00T) to about station 22+80T; from station 95+00A upstream along the grout curtain stub and left side of the inlet channel to the outlet works and across Canada de Cochiti to station 150+00A; in the low dam area from station 190+00A to 198+00A where the alignment crosses a deep arroyo; and from station 234+00A across Santa Fe River to station 257+00A to limit seepage access into the generally granular overburden or basalt bedrock foundation material. Short upstream blankets, extending 50 feet upstream from the toe of the dam, are provided in the lower dam areas, station 5+40A to 6+50A, 150+00A to 190+00A and 198+00A to 234+00A to connect the central core zone to existing surface mantle or natural blanket of finer grained overburden materials upstream of the dam. Limits of the upstream blankets are shown on Figure 1.

Cutoff and Inspection Trenches. A cutoff trench on the dam axis was excavated through the granular overburden and the full base width of the impervious core zone founded on bedrock from about station 50+00A, across the Rio Grande and outlet works areas to station 95+00A. From station 53+00A, the cutoff trench to bedrock extends upstream to a sandstone outcrop. Extension of the cutoff trench upstream from station 53+00A provides a positive cutoff along the edge of the upstream impervious fill blanket required in the area from the cutoff trench to the right abutment of the dam.

The cutoff trench across the Rio Grande stream area will effectively reduce potential sepage losses from permanent pool storage through the granular streambed alluvium.

Inspection trenches of variable width and depth were excavated into foundation overburden along the axis of the dam in areas where depths to bedrock precluded a positive cutoff. The inspection trenches increase the seepage path along foundation contact, intersect upper zones of the stratified granular foundation materials and provide a transitional connection between the impervious core zone and upstream impervious fill blankets.

Grout Curtain. A grout curtain is provided along the dam axis from about station 50+00A to 95+00A, in the cutoff trench extending upstream from 53+00A (0+00T) to about station 21+50T, and upstream from station 95+00A to the toe of the dam. The curtain reduces potential seepage losses through sandstone and basalt bedrock from permanent pool storage which is largely confined to the Rio Grande arm of the reservoir. Plan of the grout curtain is shown on Figure 13.

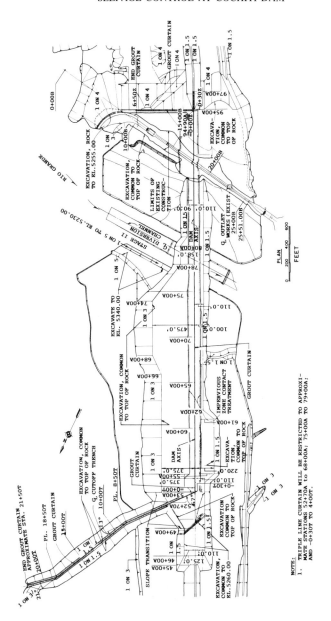

Fig. 13. - Plan of Grout Curtain (1ft=0.305m) (From Ref. 1)

Drain Holes. Areal extent of basalt within the embankment foundation area northwest of the Rio Grande is shown of Figure 14. The major portion of the basalt is upstream of the dam axis but does extend downstream of the dam axis between approximate station 52+00A to 68+00A and station 75+00A to 79+00A. Subsurface explorations indicated that downstream basalt limits were within the dam foundation area or that basalt areas under the downstream toe of the dam were severely constricted. The basalt is a potential source of high seepage volume which, if the downstream limits are blocked or severely constricted within the downstream foundation area, may cause uplift under the downstream slope of the dam.

Pressure relief in the basalt zone is provided by a series of drain holes located downstream of the grout curtain and the impervious core zone. Drain holes are spaced at 10-foot (3.05 m) centers and penetrate the total thickness of basalt to the underlying sandstone. Drain holes were drilled after curtain grouting was completed. The drain holes and the downstream drainage blanket are sized to carry the total seepage quantity estimated for the basalt zone without the grout curtain. The drain hole installation, as designed, provides a permanent, maintenance free pressure relief system.

Downstream Toe Trench. A downstream toe trench is provided in specific areas from station 20+00A to 260+32.63A. The toe trench is omitted across the Rio Grande streambed and in the outlet works area where the dam is founded on bedrock, on the abutments of Canada de Cochiti where extensive controlled grading was required, in the low dam area, and in the Santa Fe River area. The toe trench will intercept shallow foundation seepage and effectively reduce potential uplift pressure at the toe of the dam. A typical section of the toe trench is shown on Figure 15.

Downstream Blanket. A select natural pervious fill blanket, 5 feet (1.525 m) thick, is provided downstream of the embankment toe from station 6+00A to 260+32.63A. The blanket is omitted in the outlet works area where the downstream toe of the dam is on rock and in the rockfill toe area in the Santa Fe River. The exposed blanket serves as a weighted filter against foundation seepage thus providing additional protection for the toe of the dam.

Relief Wells. Relief wells are installed into foundation overburden in two separate areas along the alignment where the embankment is founded on overburden and underseepage control measures consist of an upstream impervious fill blanket supplemented by a variable depth inspection trench on the dam axis. The areas are from about station 14+00A near the right abutment to about station 55+90A and from about station 97+00A across Canada de Cochiti to station 136+00A. Relief wells are installed outside the embankment toe and at the downstream limit of the exposed drainage blanket. Relief well analyses were made using published theory and formulas for design of the pressure relief system. Assumptions, method of analysis and typical calculations are discussed and presented in reference 2.

Surface Drainage. A surface drainage system is required to intercept natural drainage onto the embankment and blankets, to

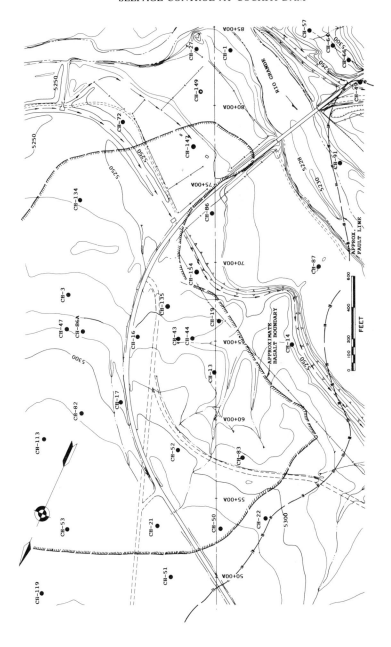

Fig. 14. - Areal Extent of Basalt (1ft=0.305m) (From Ref.3)

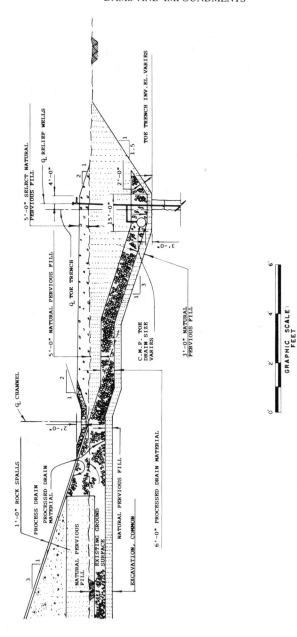

Fig. 15. - Typical Section of Toe Trench (1ft-0.305m) (From Ref. 1)

intercept and collect runoff from the steeper embankment slopes, to
control concentrations of runoff by limiting surface drainage areas
and to convey accumulated surface drainage across blanketed or other
areas to discharge points. In the downstream area, drainage channels
are required for anticipated flow from relief wells and to direct
surface runoff into the river without damage to the existing Sile main
irrigation canal and the downstream irrigated farm lands.

OPERATIONAL NOTES

Embankment Performance History. Impoundment of the permanent
pool (50,000 acre-feet)(61.5 Mm^3) began on 9 April 1975. A reservoir
elevation of 5313.00 was reached on 22 June 1975 and held until 2
October 1975. From 2 October 1975 to 18 December 1975, the reservoir
was raised to permanent pool elevation 5321.25. Prior to the flood
control storage for the 1979 spring flood season, the maximum pool
recorded since completion of the project was elevation 5331.6 (61,000
acre-feet) (75.03 Mm^3) occurring on 9 January 1976. The 1979 peak
and record flood control storage was pool elevation 5388.0 (184,430
acre-feet) (226.85 Mm^3) occurring on 21 June 1979. Field observations
and instrumentation data are discussed in detail in references 4, 5,
6, 7, and 9.

Seepage. During the first periodic inspection, May 1975,
apparent seepage was noticed in the downstream collector ditch between
station 50+00+ and station 55+00+. Picture points were established to
monitor any changes in seepage. In January 1976, minute sand boils
were observed in the same area.

Parshall Flumes. The number and location of the Parshall flumes
which were installed to monitor and measure seepage are listed below:

Flume No.	Location
1	Toe drain outlet, Sta 51+00A
2	Toe drain outlet, Sta 59+20A
3	Channel 2-6
4	Toe drain outlet, Sta 98+65A
5	Toe drain outlet, Sta 112+00A
6	Toe drain outlet, Sta 122+50A Canada de Cochiti
7	Toe drain outlet, Sta 227+00A
8	Toe drain outlet, Sta 233+30A
9	Santa Fe River
10	Toe drain outlet, Sta 253+00A

Field Observations During the Period 17-31 May 1979. Flood
control storage increased sharply from elevation 5324 (49,600 acre-
feet) (61.01 Mm^3) to elevation 5350 (88,000 acre-feet) (108.24 Mm^3).
Foundation seepage in the Rio Grande area increased with the higher
pool storage. Field observations revealed increased seepage above the
culvert headwall in channel 2-7 and into the rock-lined ditch west of
the headwall. A minor sand boil in channel 2-6 was also observed
approximately 100 feet (30.5 m) downstream from the toe of the dam

berm. Materials in this area are streambed alluvium overlain by
natural pervious fill and rock spalls. The small boil was not
considered to be a serious problem. However, it was observed
frequently during the 1979 flood control storage.

Instrumentation Data During the Period 17-31 May 1979. Upstream
piezometer readings indicated an immediate response with increase in
pool level. Downstream piezometer readings indicated little or no
response to the higher pools due to the lag time.

Field Observations During the Period 31 May 1979 - 11 June 1979.
Flood control storage increased from elevation 5350 (88,000 acre-feet)
(108.24 Mm^3) to elevation 5379 (157,000 acre-feet) (193.11 Mm^3).
Increases in seepage were noted in both the Rio Grande and Santa Fe
River areas. Seepage above the culvert headwall in channel 2-7 and
into the rock-lined ditch west of the headwall also increased with the
higher pool. The small sand boil in channel 2-6, observed during the
period of 17-31 May 1979, appeared to have stabilized. The seepage
flow was clear with minor sand movement. Seepage from the upstream
side slope of channel 2-7 and a small sand boil in channel 2-7 were
also observed.

Instrumentation Data During the Period 31 May 1979 - 11 June
1979. From the outlet works to the right end of the dam, upstream
embankment and foundation piezometer readings indicated immediate
responses to increases in pool level. Downstream piezometer readings
indicated smaller rises in water table. Piezometers installed in
granular material beneath the upstream impervious blanket at station
40+00 indicated an 85-foot (25.9 m) to 90-foot (27.4 m) head drop
through the seepage control blanket. Foundation piezometers upstream
and downstream of the grout curtain in both basalt and sandstone
foundation areas showed head drops of 17.6 feet (5.37 m) and 75.6 feet
(23.1 m) across the grout curtain as measured in foundation
piezometers FP-F1-1, FP-F2-1, FP-J1-1, and FP-J2-1, shown in figures
16 and 17. Embankment and foundation piezometer readings in the
Canade de Cochiti and Santa Fe River areas indicated little or no
response to the increase in pool levels.

Water levels in relief wells west of the outlet works increased
1 foot (0.305 m) to 2 feet (0.610 m). Water levels in relief wells
east of the outlet works and across Canada de Cochiti were small,
ranging from zero to less than one foot (0.305 m).

Seepage rates as measured in the Parshall flumes increased in
channel 2-6 and in the Santa Fe River with increasing pool levels.
Seepage measurements for the above period are presented in Table 1.

Field Observations During the Period 11-25 June 1979. During
this period, flood control storage peaked on 21 June 1979 at pool
elevation 5388.0 (184,430 acre-feet) (226.85 Mm^3). Foundation seepage
had increased in the Rio Grande and Santa Fe River areas. Seepage
measurements in Canada de Cochiti indicated very little change. New
seepage areas were observed on 25 June 1979 at the downstream toe of
the dam south of the outlet works from Station 86+00A to Station
89+00A. Seeps from the excavated sandstone faces adjacent to and

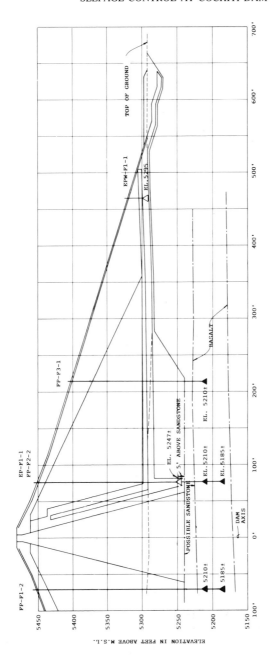

Fig. 16. - Foundation Piezometers at Station 57 + 95A (1ft=0.305m) (From Ref. 1)

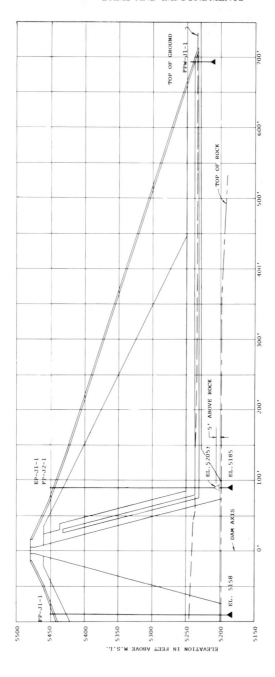

Fig. 17. – Foundation Piezometers at Station 69 + 95A (1ft=0.305m) (From Ref. 1)

TABLE 1. – Seepage Measurements During Storage of 1979 Record Pool

Date	Reservoir Elevation	Seepage in Cubic feet Per Second (m^3/sec)			
		Channel 2-6	Santa Fe River	Canada de Cochiti	Total
31 May 79	5354.8	1.36 (0.038)	2.86 (0.080)	—	4.22 (0.118)
4 Jun 79	5365.7	1.47 (0.041)	3.20 (0.090)	—	4.67 (0.131)
11 Jun 79	5379.1	1.69 (0.047)	3.44 (0.096)	—	5.13 (0.144)
18 Jun 79	5386.6	1.91 (0.053)	4.05 (0.113)	—	5.96 (0.167)
21 Jun 79	5388.0	2.03 (0.057)	3.92 (0.110)	0.05 (0.001)	6.00 (0.168)
25 Jun 79	5386.9	2.39 (0.067)	4.57 (0.128)	0.05 (0.001)	7.01 (0.196)
2 Jul 79	5384.5	2.27 (0.064)	5.25 (0.147)	0.89 (0.025)	8.41 (0.235)
9 Jul 79	5379.7	2.39 (0.067)	5.39 (0.151)	1.28 (0.036)	9.06 (0.254)
16 Jul 79	5370.1	2.27 (0.064)	5.53 (0.155)	**	7.80 (0.218)
24 Jul 79	5348.7	1.80 (0.050)	5.39 (0.151)	1.28 (0.036)	8.47 (0.237)

**Erroneous Measurement

south of the lower stilling basin were observed several weeks prior to
the above observation period. Visual observations indicated a slow
growth of these seepage areas and seepage rates during the flood
control storage.

Instrumentation Data During the Period 11-25 June 1979. The
embankment and foundation piezometers in the Rio Grande and Canada de
Cochiti areas indicated normal responses to the higher pool levels.
Foundation piezometers in the Santa Fe River area indicated small
rises in water table in the foundation basalt. The Hall foundation
piezometers FPH-B1-1 and FPH-B2-1, which were installed in the
granular foundation materials beneath the upstream impervious blanket
at Station 40+00A indicated an 85-foot (25.9 m) head drop through the
seepage control blanket. The foundation piezometers FP-F1-1, FP-F2-1,
FP-J1-1, and FP-J2-1, installed upstream and downstream of the grout
curtain in both the basalt and sandstone foundation areas indicate
head drops of 17.6 feet (5.37 m) and 75.6 feet (23.1 m), respectively,
across the grout curtain. Location of the above foundation
piezometers are shown in figures 16, 17, and 18.

Water levels in relief wells west of the outlet works rose 2
feet (0.610 m) to 9 feet (2.75 m) during the above period. Increases
in water levels in relief wells east of the outlet works and across
Canada de Cochiti varied from 0 to 8 feet (2.44 m).

Seepage rates in channel 2-6 and in Santa Fe River increased
during the above period. Seepage measurements through the Parshall
flumes are presented in Table 1.

Field Observations During the Period 25 June 1979 - 25 July
1979. Field observations were continued during the period of drawdown
from the maximum pool of record occurring on 21 June 1979. As
discussed previously, the 1979 peak storage was at pool elevation
5388.0 (184,430 acre-feet)(226.85 Mm^3). The observations during the
above period found no problem areas which could affect the stability
and safety of the dam and appurtenant project structures. During the
first two weeks of the above period, foundation seepage in the Canada
de Cochiti and Santa Fe River areas increased and seepage in the Rio
Grande area had stabilized. During the latter two weeks of the above
period, foundation seepage in the Rio Grande area decreased and in the
Canada de Cochiti and Santa Fe River areas, seepage had essentially
stabilized.

Instrumentation Data During the Period 25 June 1979 - 25 July
1979. Embankment and foundation piezometers continued to show normal
response to the peak storage and during drawdown.

Water levels in relief wells west of the outlet works and east
of the outlet works and across Canada de Cochiti continued to rise
during this period but at a much slower rate. Relief well
measurements during pool drawdown indicated a lag time of
approximately one month between maximum pool and maximum water levels
in the relief wells.

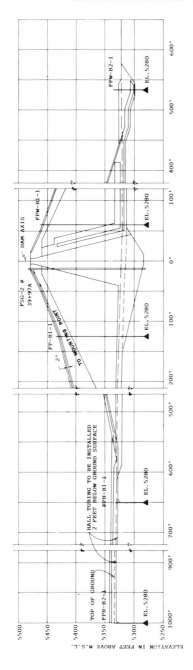

Fig. 18. – Foundation Piezometers at Station 40 + 00A (1ft=0.305m) (From Ref. 1)

Seepage rates during the first two weeks of this period increased in Canada de Cochiti and Santa Fe River areas and the seepage in channel 2-6 had essentially stabilized. Seepage rates during the latter two weeks of the above period decreased in channel 2-6 and essentially stabilized in Canada de Cochiti and Santa Fe River areas. Seepage rates measured through the Parshall flumes for the above period are presented in Table 1.

CONCLUSIONS

As previously discussed, the foundation seepage was estimated to range from 2 cu ft/sec (0.056 m^3/sec) to 8 cu ft/sec (0.224 m^3/sec) for pool elevation 5322 (Initial Permanent Pool) and from 15 cu ft/sec (0.42m^3/sec) to 36 cu ft/sec (1.01 m^3/sec) for pool elevation 5370 (Ultimate Permanent Pool). Based on seepage measurements during storage of the 1979 record pool, the maximum foundation seepage, was approximately 9 cu ft/sec (0.252 m^3/sec) resulting from the record pool at elevation 5388. As indicated above, the measured seepage quantities during storage of the 1979 record pool are less than those estimated in design.

Several factors were not included in the analyses which would tend to reduce seepage losses. In the Rio Grande area (area II), seepage through sandstone and basalt bedrock was estimated using average permeability values and no reduction in rock permeabilities were made for the beneficial effect of the grout curtain. Also in the Rio Grande arm of the reservoir, the blanketing effect from reservoir sediment deposition was not considered. Seepage estimates for flood control pools were based on complete saturation for a permanent pool at that level, whereas time duration of pool storage for flood control pools above elevation 5432 is short.

ACKNOWLEDGEMENT

The writer gratefully acknowledges Messrs. Victor Heusinger, Chief, Geotechnical Section, James O. Boardman, Chief, Design Branch, and Jasper H. Coombes, Chief, Engineering and Planning Division, U.S Army Corps of Engineers, Albuquerque District and Mr. Robert L. James, Chief, Geotechnical and Materials Branch, U.S Army Corps of Engineers, Southwestern Division, for their support during the preparation of this paper. The writer gratefully acknowledges and appreciates the efforts and expertise of Messrs. Onofre R. Gallegos and Michael Gallegos of Metro-Tec Engineering, Inc., Albuquerque, New Mexico, in the preparation of all the figures presented in this paper. The writer is grateful to Betty I. Wainright for typing the various drafts and the final copy of this paper. Last and most importantly, the writer is thankful for the love, support, patience, and understanding that his wife has provided during the preparation of this paper.

APPENDIX I - REFERENCES

1. Cochiti Dam and Reservoir, Rio Grande, New Mexico, As-Built Drawings.

2. Cochiti Dam and Reservoir, Rio Grande, New Mexico, Design Memorandum No. 9, Embankment and Conveyance Channel, Volume I, July 1967.

3. Cochiti Dam and Reservoir, Rio Grande, New Mexico, Design Memorandum No. 9, Embankment and Conveyance Channel, Appendix A, Supplemental Data on Embankment and Foundation Materials, Volume II, July 1967.

4. Cochiti Dam, Rio Grande, New Mexico, Embankment Criteria and Performance Report, prepared by Dwayne E. Lillard, April 1981.

5. Cochiti Lake, New Mexico, Rio Grande Basin, No. 1, Periodic Inspection & Continuing Evaluation of Completed Civil Works Structures, ER 1110-2-100, May 1975.

6. Cochiti Lake, New Mexico, Rio Grande Basin, No. 2, Periodic Inspection & Continuing Evaluation of Completed Civil Works Structures, ER 1110-2-100, October 1976.

7. Cochiti Lake, New Mexico, Rio Grande Basin, Periodic Inspection & Continuing Evaluation of Completed Civil Works Structures, ER 1110-2-100, Inspection Report No. 3, 9 June 1981.

8. Field Permeability Tests, Cochiti Dam, New Mexico, prepared for U.S. Army Engineer District, Albuquerque Corps of Engineers, Albuquerque, New Mexico by Fruco and Associates, St. Louis, Missouri, July 1965.

9. Monitoring Effects of Rising Water at Jemez Canyon, Cochiti, and Abiquiu Dams, New Mexico, Reports 1 through 7 for the Period 10 May 1979 through 1 August 1979, prepared for U.S. Army Engineer District, Albuquerque Corps of Engineers, Albuquerque, New Mexico, by Mr. Lewis C. Slack.

APPENDIX II - NOTATIONS

D = Thickness of pervious stratum, ft.(m)

H = Height of initial water table, ft.(m)

h_w = Head at test well, ft.(m)

$H-h_w$ = Drawdown in test well, ft.(m)

K_h = Average horizontal coefficient of permeability, fpm (m/min)

N_e = Number of Equipotential drops

N_f= Number of flow channels

Q_w= Total well flow, cu ft/min (m^3/min)

R = Radius of influence of test well, ft.(m)

r_w= Radius of test well, ft.(m)

Hydraulic Fracturing in Embankment Dams

James L. Sherard*

There is now sufficient evidence to conclude that concentrated leaks occur commonly through the impervious sections of embankment dams by hydraulic fracturing without being observed, even in dams which are not subjected to unusually large differential settlement. Usually these concentrated leaks do not cause erosion, either because the velocity is too low or because the leak discharges into an effective filter. Subsequently the leakage channel is squeezed shut by swelling or softening of the embankment material forming the walls of the crack. In the typical case no measurable leakage emerges downstream and there is no other indication that a concentrated leak developed and was subsequently sealed. This action probably occurs to some degree in most embankment dams.

Introduction

Grouting of dam foundations of both rock and soil using pressures high enough to cause the grout to flow out of the grout hole in concentrated streams by hydraulic fracturing has been understood and applied practically since before 1960. Since about 1970, it has been known that hydraulic fracturing commonly has occurred inadvertently when borings are drilled in completed embankment dam cores and the pressure in the drilling fluid at the bottom of the hole exceeds the adjacent embankment earth pressure (15)(16).

There is a substantial difference, however, between hydraulic fracturing at the bottom of boreholes by pressurized drilling fluid and hydraulic fracturing by reservoir water acting on the upstream face of the dam core causing concentrated leaks of water to enter the core. In 1970 there was no mention of this possibility in the literature and little or no discussion or recognition of the phenomenon among specialists. In a general study of dam cracking in 1972, the writer found only three pieces of literature describing experiences in which it was concluded that concentrated leaks had probably developed through embankment dams by hydraulic fracturing and had caused erosion damage or failure (9, 20, 24, 27).

Since about 1975 there has been a large change, with wide acceptance that hydraulic fracturing causes leaks through dams under certain conditions. In the last few years there has been a considerable volume of literature published on various aspects of the subject, among which the following are representative (5, 7, 8, 10, 14, 17, 25, 26).

At the present time, however, it is still not completely accepted

*Consulting Engineer, San Diego, California.

115

in the profession that concentrated leaks can develop through dams by hydraulic fracturing. Also, among specialists who accept the idea in principle, there is no widespread agreement on the conditions needed for, or the probable frequency of, concentrated leaks by hydraulic fracturing.

Evaluation of hydraulic fracturing in dams has an inherent, important difficulty. Hydraulic fractures, by definition, are only open when the water pressure acts and they are inside the dam, so it is not possible to verify the action by direct visual observation. It is necessary to infer that concentrated leaks from hydraulic fracturing develop through embankment dams by showing that there is no other reasonable cause for the leaks which are seen to develop, and by showing that the hydraulic fracturing is readily predicted by theory.

In any given case where a dam is breached or where an erosive leak develops short of failure, there is nearly always a difference of opinion among investigators regarding the origin of the initial leak causing the trouble. It is usually impossible to determine for sure what caused the initial leak because the erosion destroys any evidence which might have existed. Another common problem is that different interested parties have different vested interests in the outcome of the investigations. It is natural that the designer desires to show that the design was not at fault, that the engineer who controlled the construction wants to show that the construction supervision was done carefully, and that the owner wants to establish that someone is clearly responsible, etc.

Because of these problems, with rare exceptions, investigations of dams which have developed erosive concentrated leaks have not generally reached a definite conclusion, agreed upon by all parties, with regard to the cause of the initial leak. Frequently the majority of investigators of any given case conclude that differential settlement cracking is the only reasonable explanation for the initial leak, but there is enough smoke thrown up by the heat of the arguments presented by the minority to prevent a clear or strong conclusion.

In spite of these problems with the analysis of individual dams, as the years have gone by a fund of experience is developing which shows definite patterns. There have been quite a few records of troubles of different kinds caused by concentrated leaks through dams designed and constructed by current accepted practice. With increasing frequency the investigators have concluded that the leak developed by hydraulic fracturing.

The writer believes that there is now essentially incontrovertible evidence available to the profession supporting the conclusion that small concentrated leaks by hydraulic fracturing develop commonly in well designed and constructed dams without being recognized, even in dams not subjected to large differential settlement. This paper presents this evidence, in several main categories: (1) concentrated leaks which appear soon after the first reservoir filling through well constructed homogeneous dams with no internal filter or drain, sometimes leading to breaching failure; (2) erosive leaks through central

core dams with inadequate filters, usually occurring soon after the
first relatively rapid reservoir filling; (3) records of well constructed and designed central core dams in which nearly full reservoir
pressure was measured in piezometers at the downstream face of the core,
and in which exploratory borings in the dam made to study this condition showed conclusively that near horizontal, water-filled cracks
existed in the core; (4) FEM calculations showing that only relatively
small differential settlements will easily create stress conditions in
most dams under ordinary conditions which will allow hydraulic fracturing to occur; and (5) the discovery of "wet seams" inside impervious
dam sections, with water content higher than can be accounted for
reasonably by any other mechanism than the entry of water into an open
crack.

Before proceeding to the presentation of the experience and evidence showing hydraulic fracturing in dams is common, it is desirable
to review briefly the basic mechanics by which hydraulic fracturing can
create a leak in a dam. It has long been observed that differential
settlement open cracks appear fairly commonly on the surface of embankment dams. These open cracks do not usually extend into the embankment to depths of more than a few meters, because the internal
compressive stresses in the embankment, which increase with depth,
squeeze them shut.

In addition to open cracks the differential settlement creates
strains in the compacted embankment, changing the internal stress
distribution (sometimes called "stress transfer"). In this action
the internal pressure is increased in some locations and decreased in
others. In localized, but fairly large, zones within the dam the minor
principal stress is reduced to near zero, or to even tensile stress if
the dam has sufficient cohesion to withstand tension.

When the reservoir rises above one of these surfaces with low
stress in the impervious section of the dam, water can enter along the
surface from the upstream face in a concentrated thin layer. When the
reservoir water pressure becomes slightly greater than the embankment
stress, there is no resistance to the entry of water into the embankment in a concentrated leak. As the reservoir continues to rise, increasing the pressure in the water-filled crack, the stress conditions
in the embankment are changed, the impervious embankment material is
deformed and the crack is jacked open wider.

There has been some speculation that the reservoir water could not
penetrate the upstream face of the dam core and open a crack by hydraulic fracturing unless the reservoir water pressure were considerably higher than the earth pressure existing in the embankment on the
plane of the crack. This speculation was based on the idea that the
upstream face of the impervious embankment section could be considered
a smooth, uniformly impervious wall, in which there is no opportunity
for water penetration to start.

Experience with dam behavior has shown that this idea is not
correct. Experience shows that concentrated leaks enter cracks in the
impervious core, as the reservoir is rising, when the reservoir level

is only a meter or two above the elevation of the entrance to the crack.
This shows that no significant "excess pressure" is needed to initiate
the fracturing action. The behavior of the Wister Dam, discussed later,
where near-horizontal concentrated leaks passing with a length of more
than 200 meters through a clay dam appeared when the reservoir was being
filled for the first time and the water level was only about 4 meters
above the level of the leak, shows clearly that the water must have
entered the dam by hydraulic fracturing when the pressure was only 1 or
2 meters of water head.

      This lack of resistance to the initial penetration of water into
the upstream face of the impervious dam section is not difficult to
understand when the physical conditions are considered. All impervious
sections of embankment dams contain a multitude of cracks formed during
construction by drying, deformation under wheels of equipment and in-
completely bonded layers. These cracks are squeezed shut by the weight
of the overlying embankment but they remain nevertheless as closed
cracks in the fabric of the compacted soil which the reservoir water can
probably enter when the pressure is greater than the earth pressure,
starting the hydraulic fracturing process. After the initial penetration
of the water for a few centimeters, into these small closed cracks or
fissures at the upstream face of the impervious section, the water
pressure starts to act on the surface of least resistance (which has
total stress acting on it less than the water pressure) and the advancing
water jacks open a crack (fissure) which was not in existence before
the reservoir was raised. Probably the initial penetration of water for
a few centimeters into the upstream face of the impervious section is
facilitated by the small existing closed cracks, but the main hydraulic
fracture extending through the dam core is generally in a new crack
jacked open progressively by the penetrating water.

Experiences with Homogeneous Dams

Downstream Slope Wet Spots

      At a number of small well-constructed homogeneous dams of clayey
soil (without internal drains) over the years the writer has seen "wet
areas" appear on the downstream slope, usually in the upper portion,
soon after the first rapid reservoir. These dams usually retained small
reservoirs which filled completely for the first time in a few hours (or
a day or two) following a rainstorm. In these "wet areas" there was
generally only enough water to cause the soil surface to be completely
saturated in spots, with small surface indentations accumulating some
free water, rarely with a trickle of water running down the slope.
Within a short time (days or weeks) these wet areas on the downstream
slope often dried up completely, with no subsequent reappearance. Often
these wet spots occurred near the abutments where the dam height was
changing relatively rapidly and where tensile strains would be expected
in the upper part of the dam embankment.

      After puzzling over these temporary wet spots the writer concluded
that they must have been caused by small concentrated leaks from hydraulic fracturing. In these experiences the channel (crack width)
must have been very small with the leakage velocity far too low to
erode soil from the crack walls. In such a case, softening or swelling
of the embankment material in the crack walls probably acted to close

the leakage channels in a short time, readily explaining the observed disappearance of the wet spots.

The above experiences taken by themselves are not highly persuasive. However, they are completely consistent with the other evidence for hydraulic fracturing in homogeneous dams as discussed below.

Visible Open Cracks

Fig. 1 shows the typical pattern of open cracks seen fairly frequently in the crest of low homogeneous dams, caused by differential settlement between the portion of the dam at the abutment and in the main dam. The width of the crack at the crest depends primarily on the magnitude of the settlement and on the properties (stiffness) of the embankment, and varies quite a bit. In the writer's experience an open crack with a surface width of 5 to 10mm can easily occur as the result of the post-construction differential foundation settlement is 20 to 30 cm. Frequently the cracks are increased in width by drying and embankment shrinkage.

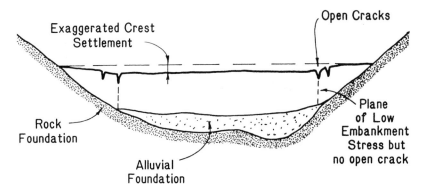

Fig. 1:   Longitudinal Section of Typical Small Dam Showing
          Common Open Cracks on the Crest and Planes of Likely
          Hydraulic Fracturing.

These cracks are commonly sealed by grouting or trenching and refilling the trench with compacted material before the reservoir is filled. The internal embankment stress acting on the transverse plane extending down below the bottom of the open crack (dotted lines, Fig. 1) is clearly not very high, and probably is tension rather than compression in many cases. A leak by hydraulic fracturing on this plane obviously could be expected.

There are many examples where low homogeneous dams, without internal drains or filters, failed rapidly by breaching (washing out of a section of the dam) immediately following the first rapid reservoir filling. There are dozens of such experiences and frequently the break occurred over the abutment, where longitudinal stretching of the dam and hydraulic fracturing is most probable (Fig. 1). For many of

them there was every reason to believe that the dams were well constructed and there was no likelihood of a concentrated leak from any
other source. Over the years the writer has had the opportunity per-
sonally to investigate a number of homogeneous dam breaching failures
in each which it was considered highly probable the initial leak
occurred in a crack which was never seen.

Stockton Creek Dam

One representative, fairly well studied, example was the breaching
failure of the Stockton Creek Dam in California immediately following
the first rapid (few hours) reservoir filling in 1950. This was a
homogeneous dam about 25 m. maximum height, founded on hard, incompressible bedrock, Fig. 2. It was well-constructed under constant,
competent engineering supervision, with the construction also inspected
by the State of California Division of Dam Safety, the strongest state
control group in the U.S.A. No open cracks were seen during or after
construction. The failure occurred when the reservoir filled rapidly
for the first time at night with no eyewitness so that it was not
known how the initial leak started. The failure breach found to
exist the next morning was about 12 m. wide at the top and was located
at the right end at a point where there was a small but near vertical
step in the sound rock abutment, where stress transfer and hydraulic
fracturing would have been most probable, Fig. 2b. After thorough
study of this experience the writer concluded that there was no other
reasonable explanation for the breaching failure than erosion of a concentrated leak which developed in a crack through the embankment from
differential settlement, although no crack was ever seen.

Since the dam foundation was relatively incompressible rock, if
hydraulic fracturing occurred the internal stress transfer must have
been caused by the differential settlement of the dam itself, which
was very low. But the material was rocky and the compacted embankment
was relatively hard and stiff, not able to adjust easily to movements
without relatively high resistance and internal stress transfer. A
more complete description with photographs is given in Ref. (16).

Oklahoma and Mississippi Failures

Another group of 14 similar breaching failures in Oklahoma and
Mississippi which occurred between 1957-70 studied by the writer are
described in Refs. (19) and (20). These were all low homogeneous clay
dams, 7 to 20 m. high, without vertical chimney drains, which were
well-constructed with competent engineering supervision and moisture-
density control. Essentially all of these failures occurred immediately
after the rapid first reservoir filling, and the initial leak appeared
at the downstream side within a few hours (or a day or two) after the
reservoir filled. In almost all cases the breach occurred at a point
along the dam length where the maximum differential settlement was expected. As discussed in detail in Ref. (20), these experiences left
little doubt that the initial leak which eroded to failure developed in
a hydraulic fracturing crack.

The main difference between this group of dams which failed by
breaching and hundreds of other similar dams built by the same engineering organization which did not fail was the fact that the failed dams

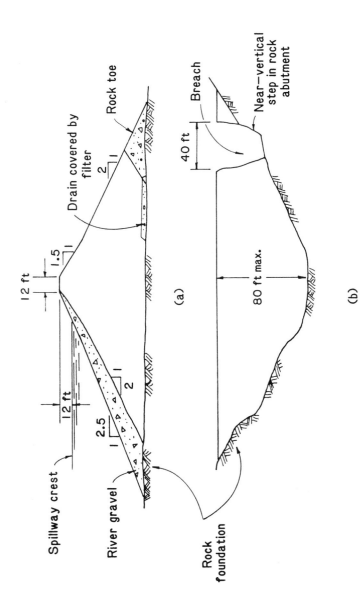

Fig. 2: Stockton Creek Dam. Cross-section (a) and Longitudinal
Section (b) Showing Breaching Failure.

were constructed in areas of highly erodible dispersive clay. For
dams of dispersive clay even a very small initial concentrated leak with
very low velocity may erode the material faster than the leakage channel
is closed off by swelling of the crack walls. For homogeneous dams constructed of nondispersive impervious soils evidently the initial leak
needs to be considerably larger and have greater velocity in order to
start erosion than for dams of dispersive clay. Therefore, everything
else being equal the likelihood of breaching failure by hydraulic
fracturing is considerably higher for homogeneous dams of dispersive
clay. This conclusion is well-supported by experience. There are
many fewer cases of breaching of this kind in homogeneous dams of
ordinary, non-dispersive soils (such as the Stockton Creek Dam failure)
than in dams of dispersive clay.

There are many other similar examples. In all these cases the
failure nearly always occurs within a very short time after the reservoir is filled rapidly, usually in one rainstorm. In these cases, the
appearance of the initial leak on the downstream slope within hours
after the reservoir rises above the elevation of the leak leaves no
doubt that the water was traveling through the dam as a concentrated
leak. The typical dam embankment is far too impervious to allow water
to travel completely through the dam in such a short time other than in
a concentrated leak.

Wister Dam

The near failure by piping of the Wister Dam in Oklahoma in 1949
is among the more important case histories in this category, and the
significance of the experience has not been widely recognized (3, 16,
23). The Wister Dam experience was very similar to the 14 failures in
Oklahoma and Mississippi described above, but occurred in a much more
important dam of the U. S. Army Corps of Engineers. The main details
were generally as follows: homogeneous clay dam (no vertical drain),
30 m. high, 1800 m. long, with reservoir volume of about 500 million m^3,
well-constructed by experienced engineers during 1946-48, Fig. 3.
In January 1949 a large storm caused the water to rise rapidly for the
first time to a maximum depth of about 18 meters on January 28, 1949.
Then, with the large bottom outlets open, the reservoir water level
started to drop slowly. On January 30, seepage carrying eroded
embankment material was first noticed on the downstream slope with
estimated quantity of about 150 l/sec. This increased to more than
500 l/sec., in the following several days, always very dirty. By February 3, with bottom outlets discharging at full capacity the reservoir level had dropped about 4 m. (maximum water depth about 14 m.)
exposing several erosion tunnels with diameter of about 0.6m entering
the upstream slope, all at about the same elevation, over a length of
about 90 meters. Subsequently, the reservoir water level was lowered
below the tunnels and the dam was saved. Studies with dyes showed that
the average leakage water travelled a distance of about 225 m. through
the dam on a gradient of about 1:50, in about 13 minutes. The leaks
travelled diagonally through the dam on a path nearly parallel with the
original river bed, through a length of the embankment which had been
constructed last as a closure section (Fig. 3).

The results of thorough studies made of the cause of the Wister

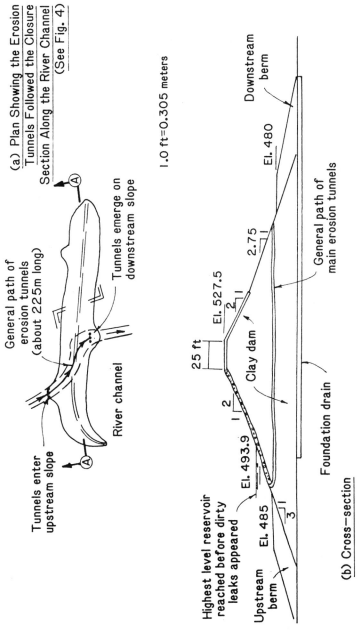

(a) Plan Showing the Erosion Tunnels Followed the Closure Section Along the River Channel (See Fig. 4)

Tunnels enter upstream slope

General path of erosion tunnels (about 225m long)

Tunnels emerge on downstream slope

River channel

1.0 ft = 0.305 meters

Downstream berm

El. 480

General path of main erosion tunnels

2.75

El. 527.5

25 ft

Clay dam

Foundation drain

2

El. 493.9

Highest level reservoir reached before dirty leaks appeared

El. 485

3

Upstream berm

(b) Cross-section

Fig. 3: Wister Dam Cross-section and Plan Showing Location of Erosion Tunnels from Hydraulic Fracturing.

Dam leakage by experienced specialized engineers were inconclusive. Finally, it was generally agreed that the initial leak must have been travelling in a differential settlement crack, even though the measured foundation settlements were not large (about 25 centimeters) and no cracks were seen. It was considered beyond the realm of possibility that the near horizontal concentrated leak entering the dam with only 4 m. head could have been caused by a poorly compacted layer or any other conceivable cause but a crack. Soon after the event, Casagrande wrote (4), "I am seriously disturbed by this case, because it demonstrates that clay which is compacted at optimum water content does not necessarily possess the ability to follow even small differential settlements without cracking."

At the time of the Wister Dam experience studied about 30 years ago, the investigators were thinking only of cracks which were open before the reservoir was filled. There was no consideration of the phenomenon of hydraulic fracturing. Later it was demonstrated conclusively that the Wister Dam was constructed of highly erodible dispersive clay, making the experience easier to understand (16)(20). During the several days when the dirty leaks were eroding the embankment material, the dam crest settled about 5 cm. over the length of the closure sections, evidently as the result of the removal of eroded clay.

The writer believes that near failure of the Wister Dam by piping is one of the more important single experiences to guide present thinking about the evolution of leakage control design measures. This was not a dam retaining a small farm pond, built with methods which could be reasonably questioned as being inadequate. This was a major dam designed and built under the supervision of the U. S. Army Corps of Engineers, one of the most experienced organizations in this type of work. All the facts were documented. The dam was well built by conventional methods from silty clay near optimum water content. The differential settlement was not high. The path of the concentrated leak was definitely determined. The entire drama was observed by experienced engineers.

The writer strongly believes that there can be no reasonable doubt that a concentrated leak developed by hydraulic fracturing completely through the dam for a distance of more than 200 m. on a near horizontal plane. This must have been caused by arching of the portion of the dam built last in the closure section between the two previously completed embankment sections on both sides of the river. The maximum total measured settlement of the alluvial foundation on the left side of the river (Fig. 4) was only about 25 cm. The details of Wister Dam experience are practically identical to the details of the erosion and breaching problems in the 15 small dams in Oklahoma and Mississippi described above (19, 20).

Probably the Wister Dam problem would not have developed if the embankment had not been dispersive clay. This conclusion is based on the knowledge that there are a great many homogeneous dams constructed of nondispersive clay which have been subjected to much more severe conditions of differential settlement than the Wister Dam which have not developed erosive concentrated leaks. Probably if the Wister Dam

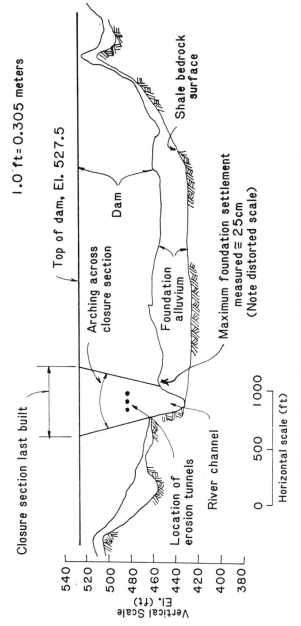

Fig. 4: Wister Dam. Longitudinal Section A-A (Fig. 3) Showing
Location of Erosion Tunnels with Respect to the Closure Section.

had been constructed of nondispersive clay the same initial leak would have developed by hydraulic fracturing, since the stress condition in the embankment needed to allow hydraulic fracturing has nothing to do with the erodibility or dispersion potential of the embankment material. The initial leakage channel was probably very narrow and the initial velocity insufficient to erode ordinary nondispersive clay. Therefore, if the initial leak had not eroded, the embankment material forming the walls of the leakage channel probably would have squeezed shut by swelling, and there would have been no indication to alert engineers at the site that hydraulic fracturing had occurred.

In summary in this section, the concentrated leak and erosion tunnels 225 m. long under 50:1 gradient, in Wister Dam is a vitally significant experience from the standpoint of improved understanding of embankment dam behavior. The importance of this experience has not been widely recognized. It shows beyond all reasonable doubt that hydraulic fracturing occurred on a path over 200 meters long when the reservoir water pressure acting on the upstream face of the dam which initiated the fracturing was not more than 4 m. of head.

The Wister Dam experience shows the development of hydraulic fracturing on a near horizontal plane caused by the tendency of the top of the dam to create a horizontal arch, Fig. 4, in the portion of the dam built last in the river closure section. The Wister Dam embankment was constructed of a typical fine-grained, medium plastic clay (CL) at average water content near Standard Proctor Optimum. The embankment was not hard and stiff as in the case of the Stockton Creek Dam described above, but rather more deformable than the average impervious dam zone. The experience shows how little settlement is needed even in a relatively deformable embankment, to cause enough stress transfer to allow hydraulic fracturing.

Experiences with Central Core Dams

The experiences described in the previous section show hydraulic fracturing in relatively low homogeneous dams where the width of the impervious section through which the leak develops is 5 to 7 times the head of water retained. For high central core dams, with rockfill or gravel shells, the impervious core width is generally less than the head of water retained and the conditions for stress transfer are much more severe. This is true because the total settlements are greater, and because internal stress transfer takes place in both directions, i.e., by transverse arching between the upstream and downstream shells and longitudinally between dam sections of different height. Therefore, if hydraulic fracturing is common in homogeneous dams with their relatively wide impervious sections, even without unusually large settlements, it should be expected even more commonly in the relatively thinner cores of central core dams. The writer believes that there is ample evidence available at the present time from records of dam performance to support this conclusion, in two main categories: (1) dams with inadequate filters which have developed erosive leaks through the core on the first reservoir filling, and (2) dams without erosive leaks in which exploratory borings in the core have encountered water-filled cracks.

Starting as early as 30 years ago a few engineers made measurements on dams leading to serious concern that "silo action" in zoned dams could reduce the total pressure acting on horizontal planes through the core to dangerously low values, or even create open cracks on horizontal planes. One interesting early example was at the U. S. Army Engineers' John Martin Dam where total pressure cells installed during construction in 1942 measured pressures on horizontal planes in the core which were less than 30 percent of the weight of column of embankment above (13).

Similar low total pressures were measured in two thin core dams in Sweden about the same time (12). In one of these Swedish dams it was estimated that the post-construction crest settlement should have been about 60 cm., and the actual measured settlement was only about 5 cm. It was the opinion of the engineers that the thin central core had probably arched between the less compressible upstream and downstream shells and that an unseen horizontal crack or series of cracks with a combined width of the order of 30 cm. could exist through the core. This hypothesis created enough concern to cause the engineers to drive a row of steel sheet piling through the core before the reservoir was filled for the first time (11).

Piping in Dams with Coarse Broadly Graded Cores

There have been more than 15 important central core dams, designed and constructed according to good current practice, in which concentrated erosive leaks have emerged suddenly at the downstream toe, usually during or soon after the first reservoir filling, discussed in more detail in Ref. (18). These are dams in many parts of the world designed and built by different experienced engineers. Most of these have been dams in which the core material was a coarse, broadly graded mixture of gravel, sand and clayey or silty fines, with many being glacial moraines. For core materials of this type, there was no general agreement on how to apply our current accepted filter design criteria, with the result that most of these dams with erosive leaks had been provided with coarse downstream filters, commonly unprocessed crushed rock or coarse alluvial gravels. These coarse filters inevitably segregate badly during construction at the core-filter interface so that they offer practically no filter action for the fine-grained portion of the material. As discussed in a companion paper (21), recent research shows that in order for a filter to be effective in sealing concentrated leaks through such a core material it must contain sufficient fine to medium sand to give a D_{15} size of 0.7 mm or smaller. Therefore, on the basis of this new understanding of filters, the coarse, segregated filters provided at the typical dam which developed erosive concentrated leaks could not be expected to seal the leaks.

In most of these cases the results of the investigations of the cause of the initial leak were inconclusive. In some of these dams extensive investigations led to the conclusion by different groups of experienced specialists that hydraulic fracturing was the only likely cause (9, 24, 25, 27). Since publication of the initial summary of this experience (18) the writer has had the opportunity to be part of the investigating team studying several other almost identical experiences, concluding that it is overwhelmingly probable that the initial leak causing the troubles in this group of dams was usually caused by hydraulic fracturing. The evidence for this conclusion consists primarily of the following elements: (1) the erosive leaks have usually occurred in conditions where low internal embankment stresses would be especially likely, such as in dams with unusually narrow, vertical cores and in locations along the length of the dam where the height is changing rapidly; (2) the initial leak appeared downstream almost immediately following the first reservoir filling, indicating that there had to be a concentrated leak through the core (there wasn't time for general seepage through the voids of the core material to have caused the leak by progressive backward piping); (3) calculations of the stress conditions (FEM) show that the low stress conditions needed for hydraulic fracturing are generally present; (4) there is no other tenable hypothesis to explain the development of the initial leaks.

In the group/of damaged dams described in this section above, the initial leak was able to erode the core because the downstream filter was too coarse to prevent erosion of the broadly graded impervious core material. In the world today there are many dozens of completely successful dams with the same type of core material and designs very similar to those of the group of damaged dams, except they had generally finer downstream filters. Undoubtedly initial leaks by hydraulic fracturing developed in many of the successful dams without being known because the filter successfully prevented erosion. The development of initial leaks by hydraulic fracturing is not limited to dams with inadequate filters.

Discovery of Water-Filled Cracks in Dam Cores

There have been a number of experiences in which piezometers in the upper half of dams located near the downstream edge of central cores have registered full, or nearly full, reservoir pressure.

Manacouagan 3 Dam: One of the most instructive of these is the experience at the Manicouagan 3 Dam (5), in Quebec, with main details as follows:

1. A central core dam with gravel shells, about 105 m. high and 400 m. long, and relatively thick vertical core of cohesionless, sandy silt, completed in 1975.

2. On the first relatively rapid reservoir filling, piezometers at the downstream face of the core registered immediately nearly full reservoir pressure, and continued to show the same pressure as the reservoir was kept nearly full for several years. Many piezometers on four instrumented sections all showed the same pattern of measurement so that it was apparent that the situation was not isolated, but a general condition.

3. To study this situation three special borings, about 75 m. deep, were drilled through the core near the downstream edge of the crest using casing without drilling fluid (mud or water), removing cuttings with compressed air. The casing was driven progressively in increments of about 2 to 4 m. At each stage in the casing descent a hole in the core was drilled 1.5 to 3.0 meters below the casing bottom and a waiting period of 1 to 2 hours was allowed during which the rising water level and elapsed time were recorded. Generally no

measurable inflow of water entered because the embankment was sufficiently impervious to limit the seepage inflow in two hours to a
negligible quantity, but at certain levels water entered rapidly. For
example, in the first boring drilled no water entered until the hole
had reached a depth of about 21 m., at which time water rose in the
casing about 11 m. in 60 minutes. Below this depth for some distance
there was again no inflow, and then again at depth about 39 m. water
rapidly entered again. The other borings had similar action.

4. There has been no measured leakage through the dam and the
piezometer readings are stable. The downstream filter is conservative,
with average D_{15} about 0.2mm. The condition is considered completely
safe.

The investigators at Manicouagan 3 concluded that "The presence of
a set of cracks (filled with loose muddy material) or decompressed
layers distributed erratically, constitutes the main cause of the high
pore pressure transmission from upstream to the downstream part of the
main dam's core ... can be attributed to arching of the dam's embankment and differential settlement of its various zones ...". (5)

The writer believes that the Manicouagan 3 Dam experience must be
recognized as providing basic proof that concentrated leaks from hydraulic fracturing develop through central core dams designed and constructed according to good current practice. The experience also
shows that hydraulic fracturing can develop in embankments of cohesionless sandy silt.

Manicouagan 3 Dam was built and the problem later studied and reported by experienced engineers, with the assistance of an eminent
Board of Consultants. The design would be generally considered very
conservative. The central core was relatively wide for this kind of
dam, about 75% of the water head. The investigations left no doubt
that there were thin cracks through the core: there had to be at least
thin open cracks to provide the needed water supply to allow the water
to rise in the casing 11 m. in 60 minutes.

Because the core material was a cohesionless sandy silt it is
probable that the roof (or hanging wall) of the crack collapsed and
the sandy silt fell down into the space being jacked open by the increasing water pressure generated by the rising reservoir. This action
would create a thin near horizontal layer which is filled or nearly
filled with very loose saturated sandy silt. The effective stress in
this layer must be essentially zero and the water pressure, about
equal to reservoir head, is the total embankment stress acting on the
plane of the layer. From the borings made in the core it was not
possible to get undisturbed samples of the embankment in the vicinity
of the water-bearing cracks, but the investigators definitely developed
the idea that there were "soft and permeable layers...composed of loose
muddy material..."(5).

As the reservoir level rose above the elevation of the piezometers
the downstream piezometers started to record nearly full reservoir
pressure with practically no time lag. This shows that the crack
developed rapidly. It was impossible for the measured pressure at the

downstream edge of the thick impervious core to have responded so
rapidly by seepage of water through the pores of the compacted core
material.

El Guapo Dam: There has been another very similar experience at
the El Guapo Dam, a central clay core, gravel-fill dam, about 65 m.
high, completed in 1978 in Venezuela, designed by experienced engineers
and well-constructed by good modern practice (6). At the El Guapo Dam
some of the piezometers near the downstream face of the clay core at
about midheight of the dam registered pressure heads only slightly less
than the reservoir level. Special borings were put down in the
suspect area, using drilling techniques similar to those at Manicouagan
3 Dam (without drilling water or mud). In one of these borings, pe-
etrating the clay core near the downstream face, water rose in the
casing fairly rapidly, stabilizing at about the pressure head measured
by the nearby piezometers.

This experience also can only be reasonably explained by a concentrated leak. This dam was of conservative design with excellent
clayey core material and sandy gravel shells. The total settlements
of the foundation (about 10 meters of dense gravel overlying
bedrock) was small and there was no tendency for unusual differential
settlement.

"Wet Seams" at Yard's Creek Dam

The experience of the Yard's Creek Dam, New Jersey, constructed
1963-65, was representative of the erosion damages in the group of
dams with cores of coarse broadly graded soils described in the last
section. This is a long, low rockfill dam with central vertical core of
broadly graded glacial moraine, with maximum height about 25 meters,
length about 2500 m., and with the earth core founded on unweathered,
impervious, hard quartzite bedrock, Fig. 5. The dam forms the upper
reservoir for a hydroelectric pumped storage project. The dam was
designed and constructed under the supervision of experienced
engineers.

During the first rapid reservoir filling very dirty leakage
emerged abruptly at the downstream toe at one location over a length
of about 100 m., reaching a maximum rate of about 100 l/sec. The experience is described in detail in Ref. (16), pp. 317-23. The problem was thoroughly investigated with generally inconclusive results
as was typical for many of these investigations, because it is impossible to determine for sure what caused the initial leak, and because different memebrs of the investigating teams had different
vested interests in the outcome of the investigations.

During the Yard's Creek investigation the writer had the rare
opportunity to enter test pits in the core of the dam, put down to the
bedrock foundation, in locations where there was no leakage emerging
at the toe. In these, some peculiar "wet seams" were observed.

As described in Ref. (16), p. 320, the glacial moraine core
generally appeared uniformly well-compacted, homogeneous and highly
impervious. In the typical compacted core exposed in the walls of the

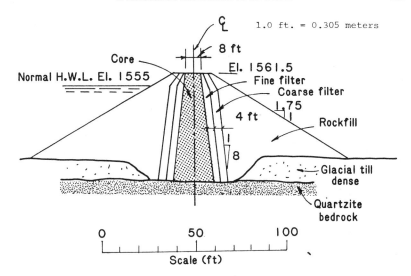

Fig. 5: Yard's Creek Upper Reservoir Dam
Cross Section

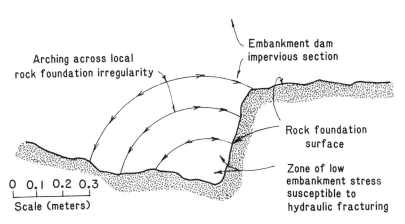

Fig 6: Zones of Low Embankment Stress Caused by Localized
Arching Adjacent to Small Irregularities in Foundation
Rock Allow Hydraulic Fracturing.

test pits, it was difficult to push in a sharp pointed steel rod more
than a few millimeters. However, there were occasional thin layers,
one to two meters apart which were much softer, obviously at a considerably higher water content than the compaction water content, into
which it was possible to push the steel rod easily. These layers were
similar in appearance to the rest of the embankment except for the soft,
wet consistency. They appeared impervious and there was no evidence
that concentrated leaks had caused any erosion.

These thin horizontal "wet seams" were obviously much softer and
wetter than could be explained by the normal action of water seeping
through the pore voids in the core and displacing the air. A specimen of soil of this kind when well compacted in the laboratory is
relatively stiff and hard, and contains about 3 or 4 percent air. If
a specimen is saturated in the laboratory at near constant volume, as
by backpressuring in a triaxial test, it remains stiff and hard. The
softening of layers in the dam must be caused by the entrance of a
significant quantity of new water associated with a near complete
elimination of the effective stress acting in the soil, as would occur
adjacent to a waterfilled, open crack.

The origin of these "wet seams" was not understood at the time
but the writer now believes, supported by the other evidence subsequently accumulated as discussed herein, that they were caused by
hydraulic fracturing, by actions approximately as described below.

Stage 1: Arching of the Core and Stress Transfer Before Reservoir
          Filling.

During and after construction the more compressible central core
settles with respect to the upstream and downstream filters, causing
arching and stress transfer. At the time the reservoir begins to
fill, the total vertical stress acting on some near horizontal planes
through the core is reduced to nearly zero, probably to the order of
1 to 2 m. of water head.

Stage 2: Initial Thin Hydraulic Fracture.

When the water rises rapidly against the upstream face of the core
during the first reservoir filling, at a certain horizontal plane the
water pressure acting on the upstream face exceeds the total earth
pressure acting on the plane. At this time the water from the reservoir is able to enter (hydraulic fracture) the core in a thin concentrated leakage channel (crack) on the near horizontal plane of low
stress. At the time of initial penetration the pressure in the water
is only slightly greater than the previous total stress on the plane,
so that there is no significant change in the overall stress conditions
in the core. The width of the initial crack is so small that the
velocity is too low to cause erosion. The fact that the water pressure
in the concentrated leak is higher than the pore water pressure in the
immediately adjacent embankment creates seepage from the crack into the
embankment, acting to hold the crack open temporarily.

Stage 3: Increasing Pressure Jacks Crack Open Wider.

As the reservoir level continues to rise, the pressure in the concentrated leak also rises, increasing the total stress on the horizontal plane, by 10-ton/m^2 for each 10 m. of rise in the reservoir level.
This increase in pressure on the near horizontal plane of the crack
causes a change in the stress conditions in the core and an increase in
the width of the crack. Rough computations using reasonable assumptions for the deformation modulus of the core, and the typical dimensions of a central core dam of moderate height, show that if only a
single crack developed, it could be jacked open to a width of the order
of 5 to 10 cm. by a pressure increase of 10 to 20 m. of water head. If
several parallel fractures developed, this computed width would be the
sum of the individual crack widths. At Yard's Creek Dam, where the
depth of water at the level of the several "wet seams" was only of the
5 to 7 m., the calculated aggregate width of cracks computed in this
way would be about 1.0 cm. During this action the water pressure in
the crack is always equal to the total embankment stress and the
effective stress in the embankment material in the crack walls is zero.

When the near horizontal crack is jacked open to any significant
width the roof must inevitably soften and collapse, tending to fill
the space created with chunks of softened embankment material, so that
the resulting condition would be a horizontal zone partially filled with
loosened embankment material surrounded by free water. The width of
such a zone depends on the dimensions of the dam and also on the distance between hydraulic fractures, when more than one develops.

Subsequently two alternate actions can occur. First, the eroded
material carried to the face of the downstream filter may seal it,
causing the water pressure in the crack for the full width of the core
to be nearly equal to the reservoir head, so that the entire head loss
occurs through a few centimeters of choked filter face. This is the
action which must have occurred at Manicouagan 3 and El Guapo Dams. In
the second alternative, the average velocity of flow in the collapsed
zone which forms along the initial leak on the horizontal plane may be
relatively low (a few centimeters/sec.) with no significant erosion or
tendency for sealing the downstream filter. In this case the head loss
will occur in hydraulic friction, and the water pressure in the crack
will vary approximately linearly across the core from upstream to downstream. Apparently this second condition is the more common situation
which develops in dams, since the high pressures measured in downstream
piezometers (as at Manicouagan 3 and El Guapo Dams) are not common.

Stage 4: Action When the Reservoir Lowers Again.

After the reservoir is lowered again below the crack level, the
water pressure in the concentrated leak reduces to zero (gage pressure),
causing the crack to squeeze shut again. Some of the free water in the
former concentrated leak is forced out, but most of it is trapped inside
the core. This trapped water and the chunks of loosened embankment
material which had fallen into the previously open crack are squeezed
together, finally creating the soft "wet seams" seen in the Yard's
Creek Dam test pits.

Only some action of this kind can result in such a soft, wet layer being found deep inside a well-constructed dam. There has to be a significant amount of water entering from the reservoir and there has to be a cavity created (open crack) in the embankment to accommodate this free water.

"Wet Seams" at Other Dams

If the mechanics of hydraulic fracturing described above are generally valid, "wet seams" similar to those seen in the Yard's Creek Dam core should develop fairly commonly in many dams. It is relatively rarely that engineers have the opportunity to examine the inside of the impervious core of a dam in test pits after the reservoir has been in operation. It is unlikely that occasional exploratory borings put down in dam cores would reliably identify wet seams. Therefore, if "wet seams" exist fairly commonly in certain locations in a significant fraction of our dams, it would not be known.

If it had not been for the investigation of the leakage problem at Yard's Creek Dam there would have been no opportunity to see the "wet seams". And in a career of specialized engineering practice devoted to embankment dams, the writer has not had many other similar opportunities.

The writer believes that it is probable that "wet seams" originating from the general action described above are not uncommon. It is impossible to prove this, but there is evidence to support the hypothesis.

The first type of evidence is from the experiences such as those of Manicouagan 3 and El Guapo Dams, described earlier. In these dams it was demonstrated conclusively that there were open water-filled cracks in the core. These cracks are in contact with the reservoir and carry nearly full reservoir pressure to the downstream side of the core. It is inconceivable that these water-bearing cracks remain completely open. The embankment material forming the roof of the crack, with zero effective stress, must collapse and fall down into the temporarily opened crack. The observation made by the investigators at the Manicouagan 3 Dam, describing layers of "loose muddy material" (5) are consistent with this hypothesis.

A second piece of evidence is the extensive series of "wet seams" discovered in the remaining intact portion of the embankment of the Teton Dam after the failure (7, 14). These are on more or less horizontal planes and are readily explained as being caused by hydraulic fracturing of the general type which occurred at Wister Dam, i.e., arching of the upper part of the dam in the longitudinal direction between the rock abutments, creating zones in the lower part where there are low total stresses on near horizontal planes.

When the Teton "wet seams" were exposed in the walls of exploratory excavations made into the dam embankment, small streams of free water dribbled out of the wet seam in the impervious compacted embankment. The time that this occurred was about 12 months after the dam failure, during all of which time the "wet seam" had not only not consolidated under the weight of the overlying embankment, it still had a lense of free water in it which would flow out when the wet seam was exposed in the exploratory excavation. This surprising behavior must be explained by the fact that a substantial volume of free water in a concentrated leak was trapped inside the dam at the time of the breaching failure.

Evidence from FEM Computations

The finite element method when applied to embankments is still in the developmental stage, in the sense that there are still doubts about how reliably embankment stresses can be predicted by the calculations. But the calculation probably does give a rough approximation of the general stress distribution and the average values. Calculations have been made to study the influence of moderate settlements on the stress distribution and the average values. These confirm generally that only relatively small differential settlements are sufficient to create large zones within dam embankments where the minimum principal stress is below the values which permit hydraulic fracturing.

As an example, in 1972 (20) some FEM calculations were made for the writer by Drs. J. M. Duncan and G. Lefebvre at the University of California (Berkeley) of a 2-dimensional mathematical model simulating one end of a low homogeneous dam of compacted clay. The end of the dam was assumed to be on an incompressible rock foundation and the post-construction foundation settlement about 150 feet away was assumed at about 30 to 50 cm. The calculations showed that even this moderate settlement created a large embankment zone near the abutment where the stresses on vertical transverse planes were in tension, and where hydraulic fracturing would be predicted, confirming the cracking commonly seen in dams in this area, Fig. 1.

In another interesting study Kulhawy and Gurtowski (10) showed that the computed stresses in the central core were generally conducive to hydraulic fracturing, even assuming: a 2-dimensional condition, an incompressible foundation and uniform properties of each embankment zone from top to bottom. Clearly similar calculations including such actual conditions as steep rock abutments, compressible foundations, and impervious cores with different compressibility at different elevations (constructed at different times with different borrow at different average water content) would obviously show much larger zones with lower internal stresses developing commonly.

In summary, embankment stress computations give clear support for the conclusion that hydraulic fracturing should be expected to be common even without unusually large settlements. The calculations show that impervious zones in embankment dams are commonly sufficiently stiff and rigid such that only small strains imposed from the outside will cause large redistributions of stress within the dam, frequently with compression changed to tensile stress.

Hydraulic Fracture Leaks Along Rock Foundation Surfaces

There is much evidence for the conclusion that small concentrated leaks commonly develop along the interface between embankment dam impervious cores and rock foundations and concrete sturctures (conduits, etc.). There are many experiences in which erosion tunnels have developed through small dams just at the foundation contact or directly adjacent to conduits passing through the base of the dam.

Probably in many cases small concentrated leaks can pass along the contact because the total pressure acting between the core and the rock foundation (or concrete conduit) is less than the reservoir water pressure (hydraulic fracturing). In one case the writer had the opportunity to see water travelling at a rock foundation contact. The test pits put down through the core at the Yard's Creek Dam in lengths of the dam with no trouble described above, were carried down to the hard rock foundation, which was cleaned off for inspection. During these studies, with the test pits still open, the reservoir level was occasionally raised 2 or 3 meters above the bottom of the pits. At this time water entered the pits in small concentrated streams, completely clear, at the rock surface with a total volume of a few l/min.

From these observations it must be concluded that the pressure existing between the dam core and the rock surface at the Yard's Creek Dam along the leakage path must have been less than 2 or 3 meters of water head. This may have been either general low pressure over the entire area caused by arching of the core between the less compressible filters, Fig. 5, or localized areas of low contact pressure caused by smaller scale arching between local irregularities in the rock surface as shown schematically in Fig. 6. Although impossible to prove or be completely confident, the writer believes it is nearly sure that small localized leaks by hydraulic fracturing of the type shown in Fig. 6 occur commonly in many dams.

At quite a few dams with leakage troubles, steps or irregularities in the rock foundation have been suspected as the most probable cause of the initial leak. One example is the Stockton Creek Dam problem discussed above, where the breach occurred at the exact location of a near vertical cliff in the hard rock abutment, about 6 m. high.

Another example is the problem of the Yard's Creek Dam, described above: erosive leakage occurred only in one section of length of the long, low dam, and the surface of the rock foundation under the core in this length was very irregular with 1 to 2 meters high near vertical steps (Ref. 16 shows photographs). Over most of the rest of the length of the dam, where no erosive leakage occurred, the rock surface under the core was more regular or smooth.

Influence of Drying and Shrinkage

Hydraulic fracturing is normally made possible by differential settlement and resulting development of zones of low internal total embankment pressures by stress transfer. Low internal embankment pressures can also be caused by drying and shrinkage, both during and after construction, especially for low dams of clay in areas of arid climate. There are many cases of erosive leaks through low dams attributed to drying cracks. In some cases these cracks may be open cracks, but more commonly they are opened by the reservoir water pressure.

A spectacular example was given by the performance of the long, low La Escondida Dam, built with homogeneous section of dispersive clay with length of 2400 meters. About 50 separate individual erosion tunnels developed at the same time in a few hours after the reservoir filled in 1972 (1). No open drying cracks were seen, but the development of so many initial leaks simultaneously must heav been caused by low internal stresses from drying and shrinkage: no other explanation is possible.

Drying and shrinkage was considered to be a contributary factor to the hydraulic fracturing of the 14 dams of dispersive clay in Oklahoma and Mississippi, described above herein. This is discussed in more detail in Ref. 20, pp. 661-665.

Influence of High Construction Pore Pressures

In dams which develop pore pressures during construction, from consolidation under the weight of overlying embankment, which are higher than the subsequent reservoir pressure, hydraulic fracturing cannot occur. In such dams the minimum principal stress always exceeds the construction pore pressures which are too high to allow hydraulic fracturing.

Relationship to Earthquake Resistance

This paper is devoted essentially entirely to the conclusion that both experience with dam behavior and theory require the dam design to assume that concentrated leaks by stress transfer and hydraulic fracturing develop through impervious sections of most dams, even without exceptional settlement. The initial leak generally occurs during or soon after the initial reservoir filling, when the reservoir water first encounters the embankment zone of low stress.

During earthquake the embankment may be again distorted by slumping. This again causes stress transfer and a second opportunity for hydraulic fracturing, which can be more extreme (larger zones of lower stress) than the conditions existing during first reservoir filling.

Summary and Conclusions

1. Experience with dam performance and evidence from theory is presented to support the conclusion that concentrated leaks by hydraulic fracturing occur frequently without being recognized.

2. This action occurs inside the dam only after the reservoir rises and cannot be demonstrated by direct visual observation. The evidence available is nevertheless sufficiently strong that the main conclusion must be accepted: there can only be some remaining questions about the details.

3. Hydraulic fracturing is made possible by differential settlement and internal stress transfer. This occurs even in dams without unusually large settlement. The hydraulic fracturing may develop on near vertical transverse planes caused by longitudinal stretching (Fig. 1) on near horizontal planes caused by horizontal arching (Fig. 4), or a more complex combination of actions.

4. These conclusions are based primarily on analyses of dams

which have developed concentrated leaks, many of which leaks cannot be
otherwise explained reasonably. The conclusions are also unequivocally
supported by FEM computations which show that only relatively small
settlements of the foundation, or the lower part of the dam itself, are
needed to create zones of low total stress in the dam capable of allowing hydraulic fracturing.

5. The concentrated leaks are not usually recognized because they
do not erode the impervious embankment material, and because the water
pressure in the leak (crack) across the core is generally about the
same as would be measured with piezometers if the concentrated leak did
not exist.

6. The action occurs in dams of all sizes, ranging from low
homogeneous dams to high central core dams.

7. The width of the hydraulic fracturing crack depends largely
on the pressure of the water entering the crack (and the reservoir
depth above the crack). For low dams the cracks are probably generally
very narrow, such as a small fraction of a millimeter. For high dams
calculations show that the cracks can be jacked open several centimeters.

8. In the general case the concentrated leaks do not erode the
core either because the velocity is too low or because they discharge
into a filter which effectively prevents erosion.

9. If no erosion occurs the initially open crack is filled by
swelling or softening of the adjacent embankment material, creating soft
layers ("wet seams"). After the crack is thus filled, the water
pressure distribution through the "wet seam" is generally linear from
reservoir pressure to zero (gage) pressure at the downstream end. In a
few cases, as demonstrated by the Manicouagan 3 and El Guapo Dam experiences, the hydraulic control develops at the downstream end of the
crack by sealing the filter, in which case the water in the crack for
its full length through the core has nearly full reservoir pressure.

10. Occasionally in low homogeneous dams without internal filters
the initial leak is able to start eroding and the dam fails by
progressive erosion. This happens much more easily and commonly in
dams built of highly erodible dispersive clay. Only rarely do failures
of this kind occur in low homogeneous dams of ordinary non-dispersive
impervious soil; but it is impossible to predict under what circumstances they will fail. The Stockton Creek Dam failure occurred in a
well-built homogeneous dam which had very low settlement.

11. The leak through the Wister Dam could not have developed by
any other mechanism but hydraulic fracturing. This experience is completely and reliably documented. The Wister Dam experience alone is
sufficient to support the conclusion that concentrated leaks develop
by hydraulic fracturing even in dams with low to moderate settlement.

12. Concentrated leaks along the contact between rock foundations
and impervious cores may be caused by hydraulic fracturing. The low
stress allowing fracturing may be general on the surfaces or it may be

locally developed adjacent to irregularities in the rock foundation
surface.

13. Hydraulic fracturing can occur in dams with impervious
sections of cohesionless silt as well as sections of clayey soil, as
demonstrated by the Manicouagan 3 Dam behavior.

Appendix 1 -- References.

1. Benassini, A. and Casales, V., "Discussion of Piping Failure of La
    Escondida Dam, Mexico," Specialty Conference on Performance of Earth
    and Earth-Supported Structures, ASCE, June 11-14, 1972, Vol. 3,
    pp. 135-141.

2. Bertram, G. E., "Experimental Investigation of Protective Filters,"
    Soil Mechanics Series No. 7, Graduate School of Engineering, Harvard
    University, 1940.

3. Bertram, G. E., "Experience with Seepage Control Measures in Earth
    and Rockfill Dams," 9th ICOLD, Istanbul, Vol. III, 1967, pp. 91-109.

4. Casagrande, A., "Notes on the Design of Earth Dams," Journal of the
    Boston Society of Civil Engineers, Vol. 37, Oct. 1950.

5. Dascal, O., "Peculiar Behavior of the Manicouagan 3 Dam's Core",
    International Conference on Case Histories in Geotechnical Engin-
    eering, May, 1984, University of Missouri-Rolla, Vol. II,
    pp. 561-569.

6. Flores, C. E. and DeFries, C. K., "Incidents with Earth and Rockfill
    Dams in Venezuela", (Spanish), Proceedings of the 25th Anniversary,
    Venezuelan Society of Soil Mechanics and Foundation Engineering,
    Caracas, 1983, pp. 243-251.

7. Independent Panel to Review Cause of Teton Dam Failure, "Report on
    Failure of Teton Dam," U. S. Government Printing Office, Washing-
    ton, D. C., 10401, 1976.

8. Jaworski, G. W., Duncan, J. M. and Seed, H. B., "Laboratory Study
    of Hydraulic Fracturing", Journal of the Geotechnical Engineering
    Division, ASCE, June 1981, pp. 713-731. (Discussion in Nov. 1982
    Journal.)

9. Kjaernsli, B. and Torblaa, I., "Horizontal Cracks Through the Core
    at Hyttejuvet Dam," Publication No. 80, Norwegian Geotechnical
    Institute, Oslo, Norway, 1968.

10. Kulhawy, F. H. and Gurtowsky, T. M., "Load Transfer and Hydraulic
    Fracturing in Zoned Dams," Journal of the Geotechnical Division,
    ASCE, Sept. 1976, pp. 963-974.

11. Lofquist, B., "Earth Pressure in a Thin Impervious Core," 4th ICOLD
    New Delhi, Vol. 1, pp. 99-109, 1951.

12. Lofquist, B., Discussion of Cracking in Dams, 5th ICOLD, Paris, 1955, Vol. III, pp. 21-22.

13. Penman, A. D. M., "Materials and Construction Methods for Embankment Dams and Cofferdams" General Report, Question 52, 14th ICOLD Rio de Janeiro, 1982, Vol. IV, pp. 1105-1228.

14. Seed, H. B. and Duncan, J. M., "The Teton Dam Failure -- a Retrospective Review," 10th ICSMFE, Stockholm, Vol. 4, 1981, pp. 214-238.

15. Sherard, J. L., "Loss of Water in Boreholes in Impervious Embankment Sections," Proceedings, 10th ICOLD Congress, Montreal, Vol. VI, 1970, pp. 377-381.

16. Sherard, J. L., "Embankment Dam Cracking," Embankment Dam Engineering, John Wiley and Sons, New York, 1973, pp. 272-353.

17. Sherard, J. L., Discussion of "Load Transfer and Hydraulic Fracturing in Zoned Dams," by F. H. Kulhawy and T. M. Gurtowsky, Journal of the Geotechnical Engineering Division, ASCE, July 1977, pp. 831-833.

18. Sherard, J. L., "Sinkholes in Dams of Coarse, Broadly Graded Soils," 13th ICOLD Congress, India, Vol. II, 1979, pp. 25-35.

19. Sherard, J. L., Decker, R. S. and Ryker, N. L., "Piping in Earth Dams of Dispersive Clay," Proceedings of the Specialty Conference on Performance of Earth and Earth-Supported Structures, ASCE, June 1972, pp. 589-626.

20. Sherard, J. L., Decker, R. S. and Ryker, N. L., "Hydraulic Fracturing in Low Dams of Dispersive Clay," Proceedings of the Specialty Conference on Performance of Earth and Earth-Supported Structures, ASCE, June 1972, Vol. 1, Part I, pp. 563-590.

21. Sherard, J. L. and Dunnigan, L. P., "Filters and Leakage Control in Embankment Dams, ASCE, Denver, Colorado, May 2, 1985.

22. U. S. Army Engineers, "Filter Experiments and Design Criteria," Technical Memorandum No. 3-360, Waterways Experiment Station, Vicksburg, Miss. 39180, 1953.

23. U. S. Army Engineers, "Review of Soils Design, Construction and Performance, Wister Dam, Oklahoma," Technical Report 3-508, June 1959, Waterways Experiment Station, Vicksburg, Miss.

24. Vaughan, P. R. et al, "Cracking and Erosion of the Rolled Clay Core at the Balderhead Dam," Proceedings of the 10th ICOLD Congress Montreal, Vol. 3, 1970, pp. 73-93.

25. Vaughan, P. R. and Soares, H. F., "Design of Filters for Clay Cores of Dams," Journal of the Geotechnical Engineering Division, ASCE, January 1982.

26. Widjaja, H., Duncan, J. M., and Seed, H. B., "Scale and Time Effects in Hydraulic Fracturing," Research Report to U. S. Army Engineers, Waterways Experiment Station, P. O. Box 631, Vicksburg, Miss. 39180, 1984.

27. Wood, D. M., Kjaernsli, B., and Hoeg, K., "Thoughts Concerning the Unusual Behavior of Hyttejuvet Dam," 12th ICOLD Congress, Mexico, Vol. II, 1976, pp. 391-414.

SOLDIER CREEK DAM FOUNDATION DRAINAGE

By Richard L. Wiltshire,[1] M. ASCE

ABSTRACT: A possible mode of failure for Teton Dam, piping of erodible core material through unsealed foundation joints, was also concluded as being possible at Soldier Creek Dam, Utah. Evaluations examined the quality of the constructed foundation and the characteristics of the dam embankment. A number of alternative treatments were studied and certain ones selected after modification of the dam was concluded as being necessary. Drilled foundation drains from existing or constructed abutment tunnels were constructed in addition to a few drains drilled from the downstream abutment surface. A long drain hole and pipe 595 feet (181 m) in length were constructed to carry collected seepage from the right abutment tunnel. Slotted PVC pipes wrapped with geotextile for a filter were installed in the drain holes. The drains, constructed to within 25 feet (8 m) of the grout curtain, protect as much of the core as possible. The foundation drains appear to be functioning as intended since a significant amount of seepage is intercepted, filtered, and removed from the foundation.

INTRODUCTION

The piping of silty core material into unsealed foundation joints through which seepage carried the material out beyond the confines of the dam embankment was postulated as a possible cause of the failure of Teton Dam, Idaho, in 1976 (ref. 2). Soldier Creek Dam was designed and constructed with certain characteristics which were similar to those of concern at Teton Dam. The evaluation of such dams and the design of modifications to prevent piping failures are complex problems.

The paper describes the evaluation of the dam characteristics related to piping failure, the design of dam modifications to address the potential problems, and the construction of required modifications at Soldier Creek Dam. Hopefully, this paper will help others involved in the evaluation and treatment of existing dams.

GENERAL

Soldier Creek Dam is an existing earth dam located on the Strawberry River in east-central Utah about 40 miles (64 km) west of Duchesne and about 35 miles (56 km) east of Provo, Utah. Construction of the dam began in 1970 and finished in 1973. Soldier Creek Dam inundates Strawberry Dam located 7 miles (11 km) upstream at a crest elevation of 7571 feet (2309 m), and enlarges Strawberry Reservoir

---

[1] Civil Engineer, U.S. Bureau of Reclamation, Engineering and Research Center, Embankment Dams Branch, Denver, Colorado 80225

fourfold to 1.1 x 10^6 acre-feet (1.4 x 10^9 m³). The dam, 3.4 x 10^6 yd³ (2.6 x 10^6 m³) in volume, rises 265 feet (81 m) above streambed with a crest length of 1,325 feet (404 m). The dam is a zoned embankment containing a central core, a downstream chimney drain connected to a blanket drain, and upstream and downstream shell zones (fig. 1). The upstream slope is protected by riprap above the top of inactive storage and the downstream slope is covered by a thin zone of rockfill raked from the downstream shell zone. Flood control is provided by storage above the top of the active conservation pool, by releases through the upper and lower outlet works, and by releases through Strawberry Tunnel to the Great Salt Lake Basin. The reservoir is still in first filling because of restrictions after which the reservoir will take a number of years to fill as the other Central Utah Project features come on line.

In 1977, the U.S. Department of the Interior authorized W. A. Wahler & Associates to evaluate Soldier Creek Dam (ref. 5). W. A. Wahler & Associates concluded that there could be several deficiencies which could cause serious distress or failure of the dam. They made recommendations to address and resolve the concerns raised about Soldier Creek

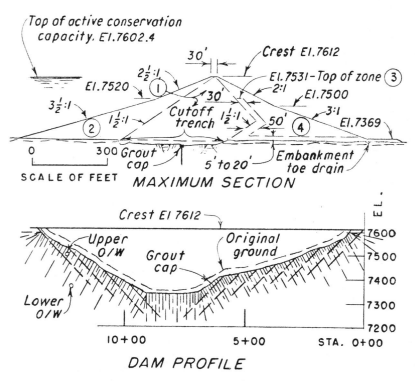

Figure 1. - Maximum Section and Profile

Dam which involved sampling and testing the dam, installation of
instrumentation, and analyses to evaluate the concerns. They also
recommended the reservoir not be raised above the pool at that time
until the concerns had been resolved.

The Bureau of Reclamation followed these recommendations by
restricting the reservoir pool to elevation 7500 feet (2288 m) and
developing a field program for sampling and testing the embankment and
foundation and for the installation of piezometers in the embankment
and foundation at two sections. Laboratory testing of the embankment
samples was performed to examine the material characteristics of
interest. The resulting information was assembled and, along with
other relevant data, was reviewed by Ralph B. Peck who had been
retained as the Bureau's consultant. His evaluation (ref. 3) and the
Bureau's own evaluation reached the same conclusion: that modification
of the dam was required to provide additional safety against potential
piping failure of the embankment.

The consultant's evaluation concluded that interception and drainage
of the seepage through near-surface foundation joints close to the
grout curtain was required. This would protect the remaining embankment core contact downstream of the drainage measures. Various schemes
for draining the foundation at Soldier Creek Dam were developed and one
approach was selected, again with Peck as consultant (ref. 4). The
selected modifications were designed by the Bureau and were constructed
in 1983 and 1984 in two contracts. The filling criteria for and monitoring of the reservoir and dam have been revised in response to the
constructed modifications and the dam safety concerns, and the filling
of Strawberry Reservoir has resumed.

EVALUATION OF THE EXISTING DAM

Many aspects of the dam's design and construction were involved in
evaluating the potential for piping failure at Soldier Creek Dam.
These aspects are divided into those concerning the foundation and
those concerning the embankment.

Foundation Factors. - The most important factor involved not sealing
bedrock joints at the zone 1 (core) contact by such methods as slush
grout or dental concrete. Another important factor was the bedrock
joint system at the site. The single row grout curtain with primary
and secondary holes was also a significant factor. Lastly, the grout
cap used for the curtain grouting was of concern.

The bedrock at Soldier Creek Dam is Green River Formation of Upper
Tertiary age and consists of lacustrine beds of sandstone, siltstone,
and shale with calcareous cementation. Some fluvial deposits of indurated, coarse grained, clean sand with small conglomerate lenses exist
within the lacustrine beds. Nearly horizontal bedding occurs distinc-
tly at one-half inch (13-mm) to 2-inch (50-mm) intervals in the shales,
at 6-inch (150-mm) to 2-foot (0.6-m) intervals in the siltstones, and
is indistinct and massive in the sandstones. Regional jointing is
spaced 6 to 30 inches (150 to 750 mm) apart, is very prominent, and is
found throughout all of the Green River Formation. The two primary
joint systems are vertical, oriented at about 45° angles to the valley
at the damsite and varied from closed to open 4 inches (100 mm) in
width. Vertical relief joints oriented parallel to the valley were
also present and varied from closed to open 2-1/2 inches (60 mm)
although some openings as much as 6 inches (150 mm) were reported after

right abutment excavation. Stinking Springs fault is located about 600 feet (180 m) beyond the right end of the dam and dips at 75° toward the west, away from the dam. The right abutment was more fractured than the left abutment bedrock, probably due to the proximity of the fault. The valley bottom was also fractured but somewhat less than the abutments. Water tests indicated the joints were tight below a depth of 50 to 70 feet (15 to 21 m) in both abutments and the valley bottom.

The foundation excavation was adjusted during construction to try to improve its quality as a seepage barrier. The lower portion of the key trench for the grout cap was widened up to three times the original 30-foot (9-m) width and was excavated 15 feet (5 m) deeper into bedrock to remove the more jointed surface rock and to provide a good foundation for the grout cap. However, some through-going near-surface joints may not have been removed which, if not grouted, could provide dangerous zigzag pattern seepage paths through the foundation.

Common practice in the design and construction of earth dams is the sealing of foundation joints, cracks, and other features with non-erodible material such as slush grout and dental concrete, especially at the core-foundation contact. At Soldier Creek Dam, a Bureau practice of relief joint grouting was utilized where joints open more than 2 inches (51 mm) received a 2-inch (51-mm) black pipe inserted into the joint. The pipe was bent to conform to the abutment shape. The embankment fill was brought 20 feet (6.1 m) higher and the pipe was grouted using a 1:1, water:cement ratio, mix after water testing the pipe to check for obstructions. This procedure was of concern for two reasons: (1) the grouted core-foundation contact was not visible to evaluate the quality of the "sealed" contact, and (2) joints less than 2 inches (51 mm) in width were not grouted. The relief joint grouting used 149 pipes and 14,529 sacks of cement in the effort to seal the core-foundation contact.

Soldier Creek Dam has a single row grout curtain which consisted of initial (primary) and secondary holes. The initial holes were inclined 30° from normal into the abutment and the secondary holes were inclined at 45° from the initial holes in the opposite direction. The base pattern for the initial holes used 160-foot (49-m) holes on 40-foot (12-m) centers split by 60-foot (18-m) holes. Initial holes started with a grout mix of 10:1 water:cement and either obtained refusal or the mix was thickened to obtain refusal. A 3:1 mix was the thickest used on the initial holes. The 181 initial grout holes totaled 13,177 feet (4,019 m) of hole which accepted 3,982 sacks of cement for 0.3 sack per foot (0.3-m) of hole average. The 13 secondary holes accepted 188 sacks in 2,894 feet (883 m) for an average take of 0.06 sack per foot (0.3-m). Secondary hole takes of up to 50 sacks were experienced and indicate that "windows" should be assumed to exist in the existing grout curtain. Such windows in the curtain, whether deep or near-surface, would be the source of concentrated seepage through the curtain into the joints downstream.

The Bureau often uses a grout cap 3 feet (1 m) wide and 3 feet (1 m) deep to strengthen the surface of the foundation for grouting purposes as at Soldier Creek Dam. Depending on the foundation treatment, this grout cap can develop very high hydraulic gradients over or beneath it. If the joints upstream and downstream of the grout cap are not filled by such as blanket grouting, it is possible that the full reservoir head might have to be dissipated going over or under the grout cap. With a head of 250 feet (76 m) in the valley bottom at Soldier Creek Dam, the gradient across such a grout cap might be 50 or so and would be excessive. Blanket grouting was performed at Soldier Creek Dam from 222 feet (68 m) upstream to 12 feet (4 m) downstream of the grout cap using 2,185 sacks of cement. This grouting usually started with a 6:1 ratio mix or thicker and achieved refusal or went to a thicker mix. These holes were often drilled at an angle to intercept joints (no pattern was used on the blanket grouting holes). Piezometers installed on either side of the group cap are indicative of the gradient across the grout cap. The instrumentation data are covered later in another section of the paper.

Summarizing the foundation factors discussed, a number of important improvements to the foundation were constructed at Soldier Creek Dam. These included removal of the most fractured surface bedrock in the key trench and the widening of the key trench in the lower portions. Note that the entire core contact surface was not thus treated. Also, the large amount of grout pumped into the bedrock should have improved its seepage barrier quality tremendously. Of the negative factors, the lack of slush grout or dental concrete sealing the surface joints at the contact with the zone 1 core is unquestionably the most critical. The relief joint grouting probably did much to improve this situation but ungrouted surface joints in the foundation must be assumed to exist because the joints were covered by the embankment during this grouting and complete sealing of the joints could not be verified. In addition, the orientation of the regional and relief joints is such that a large degree of horizontal communication through these joints must be assumed. The (clayey) shale beds restrict vertical communication. The lack of proper joint sealing at the contact with the core allows seepage erosion of core material into the foundation joints which could lead to piping failure of the dam.

Embankment Factors. - The most important embankment factor involved the erodibility of the zone 1 core material placed against the foundation and in the rest of the core. Another important factor was the gradation of the zone 1 material when considering its potential for self-healing in the event core erosion did occur.

The erodibility of a material depends primarily on its Atterberg limits and on its gradation. These characteristics varied for the zone 1, mostly because six borrow areas were utilized to construct the 1.6 x 10^6 yd^3 (1.2 x 10^6 m^3) embankment core. The preconstruction exploration and testing of these borrow areas included Atterberg limits, gradation, specific gravity, etc., to determine suitability for use within the various zones of the embankment. Table I and figure 2 present the more pertinent data for the borrow areas utilized in the core's construction. The zone 1 special compaction material placed against the foundation came from the same borrow areas. Some adequate clay was usually available in each borrow area but may have been somewhat spotty in location. The zone 1 special compaction material was to be the most clayey material available in the borrow area being worked at the time. The same six borrow areas used for the core also provided some or all of the material used in zones 2 and 4, 790,000 yd^3 (604,000 m^3) and 380,000 yd^3 (291,000 m^3), respectively. The low plasticity material was probably used in zones 2 and 4 and the better clay material when found was probably saved for zone 1.

Table I

	Borrow area source					
	A	B	C	D	F	G
LL-Ave	23%	23%	22%	23%	21%	23%
-High	32%	40%	38%	33%	30%	26%
-Low	18%	15%	16%	18%	8%	18%
PI-Ave	6%	4%	3%	5%	2%	2%
-High	13%	17%	17%	16%	10%	8%
-Low	2%	0%	0%	0%	0%	0%
No. Samples	11	160	79	55	74	13

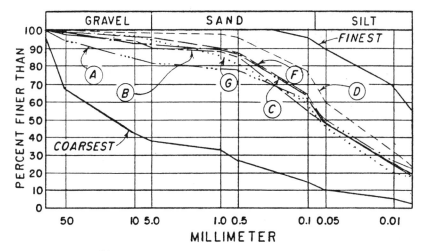

Figure 2. - Borrow Area Average Gradations

Two embankment crest drill holes were advanced through the zone 1 core during the postconstruction installation of instrumentation and were used to obtain 40 undisturbed (Denison) samples of the core. The drill holes averaged liquid limits of 24.9% and 23.0% and plasticity indexes of 7.9% and 7.4%, respectively, for 19 and 21 samples. The highest plasticity indexes were 24% in the first and 15.0% in the second hole, both values located at the contact with the foundation. These data, though not a proper statistical basis, indicate the situation may be better than thought from the borrow area data.

Another set of drill holes from the installation of instrumentation produced some very damaging information. Four holes were drilled into the foundation near the grout curtain and the zone 1 cores located adjacent to the bedrock were retrieved. Three of these zone 1 cores appeared to be good, plastic clay material but the other core was non-plastic. The evaluations of the dam concluded from this information that low-plasticity or nonplastic core material must be assumed to be located against the foundation at various locations. The evaluations also concluded that some of these locations must be assumed to be in contact with one or more of the ungrouted joints capable of conducting foundation seepage. The probabilities of these occurrences were judged to be low, but not negligible.

The erodibility of the core materials thus became an important question. The nonplastic and low-plasticity core materials could be highly erodible based on previous experience with similar materials. The question became just how erodible were the more average zone 1 materials and the more average zone 1 special compaction materials placed against the foundation. As a check, the gradation data for the drill hole samples (fig. 3) should be compared with those from the borrow areas. The drill hole sample data fall within the data from the borrow area samples. Based on the average gradation and Atterberg limits data, it was estimated that the zone 1 and zone 1 special compaction materials possessed intermediate to moderate piping resistance. Laboratory tests to evaluate the erodibility of the zone 1 materials were performed as a check on the estimated resistance.

The first laboratory test used a modified version of the Bureau's Pinhole Test apparatus (ref. 1) to determine if water running through the pinhole could erode the zone 1 material of the specimen. A number of specimens from the drill hole samples were used and three different water flow velocities were tried. No material eroded from the wall of the pinhole but these results were not considered reliable for several reasons. The specimen was small, measuring 34 mm in diameter and 38 mm long. The pinhole through the specimen was also small at 1 mm in diameter. The wide gradation of the material tested, the grain sizes involved, and the small pinhole diameter probably prevented the erosion of material in the tests.

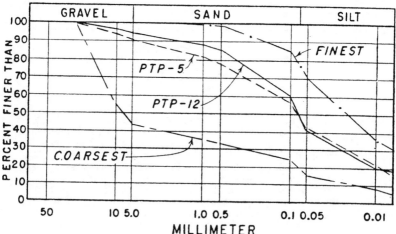

Figure 3. - Drill Hole Average Gradations

The second erosion test involved filter testing of the zone 3 material used in the drainage zones. This test used a 12-inch (305-mm) diameter permeability apparatus where the zone 3 as the filter material was placed and compacted in five lifts in the bottom and the zone 1 material to be filtered was placed and compacted in five lifts above the zone 3. The initial part of the test tried to see if water percolating through the zone 1 could erode it into the zone 3 material below. No movement of zone 1 material was noted in this part of the tests. Eight holes 1/4-inch (6-mm) in diameter were then punched through the zone 1 to the contact with the zone 3. This permitted water to run through the eight holes into the zone 3 material. Three specimens were produced using zone 1 samples from the drill holes and using zone 3 samples obtained from the borrow area. The zone 1 samples were grouped into similar Atterberg limits, and five samples per specimen were selected and used. The three zone 1 specimens had average plasticity indexes of 8.8, 10.2, and 14.4% in the erodibility testing. The zone 1 material eroded from the wall of the punched holes in the tests of all three specimens although erosion took longer to develop with the higher plasticity index materials. Erosion development was indicated by the rate of water flow through the apparatus; the flow would decrease dramatically when sufficient material eroded from the walls to plug off the hole where the material was filtered by the zone 3 below. When examined, the walls of the lowest plasticity index holes appeared to be completely eroded. The walls of the highest plasticity index holes were still shiny (from the punching of the holes) over much of the hole length. In addition, the lowest plasticity index holes were full of eroded material compared to the highest plasticity index holes which were only one-third full. These test results agree with the generally accepted rule that the higher the plasticity index, the less erodible and less pipable the soil is.

Summarizing the embankment factors discussed, a number of both good and bad conclusions were reached. The zone 1 material is erodible. Some portions of the core are probably highly erodible while other portions are much less erodible. The zone 1 special compaction material placed against the foundation probably belongs to the latter category over most of the foundation, but some areas must be assumed to contain highly erodible material. The gradation of the zone 1 material is variable but is probably sufficiently well graded on the average to help resist erosion. Such material might also be self-healing if erosion did occur.

Instrumentation Indications. - Fourteen observation wells were installed in the abutments during the dam's construction. Porous tube piezometers (32) were installed in the embankment and foundation in 1981 during the sampling instrumentation and testing program for the dam's evaluation. The piezometers were installed at two sections, stations 5+70 and 8+70, which correspond to part way up the right abutment where grout takes were largest and down at the maximum section, respectively. At both sections, piezometers were placed in the embankment at the base and in the foundation both upstream and downstream of the grout cap and curtain. This was intended to measure the head loss in both the embankment and foundation for seepage traveling over or through the grout curtain. Additional foundation piezometers (6) were installed prior to drainage construction in 1982 to monitor the effects of the drainage installation.

The piezometers at both station 5+70 and station 8+70 (fig. 4) indicate substantial head losses from the reservoir to downstream of the grout curtain with most of the loss occurring in the area of the grout curtain. The station 5+70 foundation piezometer just downstream of the grout curtain, located about 130 feet (40 m) below the maximum reservoir water surface to date which occurred in June 1984, indicated 20 feet (6 m) of pressure head at that time or only 15 percent of the initial potential head at the reservoir. The head loss between these upstream and downstream piezometers was 85 feet (26 m). The station 8+70 foundation piezometer just downstream of the grout curtain, located about 210 feet (64 m) below the same reservoir water surface, indicated 40 feet (12 m) of pressure head or only 19 percent of the initial potential head. The head loss between these upstream and downstream piezometers was 117 feet (36 m). Note that some head was lost in traveling to the upstream piezometers at both sections. The upstream and downstream piezometers at station 5+70 are 25 feet (8 m) from the grout curtain and at station 8+70 they are 50 feet (15 m) from the curtain. At both embankment sections, the indicated foundation pressure head elevation downstream of the grout curtain is below the indicated embankment pressure head elevation. At station 5+70, this foundation pressure is 10 feet (3 m) below the embankment pressure and, at station 8+70, this foundation pressure is 32 feet (10 m) below the embankment pressure. This indicates that at both locations, the embankment seepage is moving down into the foundation.

The reservoir rose 67 feet (20 m) in 1983 and 47 feet (14 m) in 1984, which was reflected in some piezometer readings. The piezometers that changed with the reservoir were those in the foundation or near the foundation in the embankment core. Thus, the head losses indicated by the piezometers over or through the grout curtain are not significantly affected by seepage sources other than the reservoir. The core and the foundation in the area of the grout cap and curtain appear to be very good barriers to seepage at the two instrumented embankment sections, and perhaps over the rest of the dam, too.

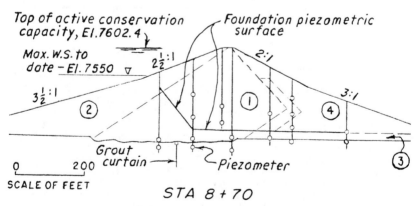

Figure 4. - Station 8+70 Foundation Piezometric Surface

Evaluation Conclusions. - The various aspects of the dam's design and construction discussed contained both good and bad findings. Included in the good factors, the most important was probably the apparent good quality of the grout curtain as a seepage barrier. Another good factor was the probable higher average plasticity index of the zone 1 special compaction material against the foundation compared to the zone 1 material, and the indications that the as-built core had better plasticity than indicated for the borrow areas. Another good characteristic of the zone 1 material was the fact that it was well graded on the average for good self-healing reaction if eroded. Included in the negative factors, the most important was the unsealed joints at the foundation contact with the embankment core. Almost as serious was the erodibility of the zone 1 material under certain seepage conditions. Another detraction was the possibility for ungrouted windows in the grout curtain.

The evaluation into the quality of the existing dam, summarized by the above most important factors, concluded that the potential for failure of the dam was low but not negligible. The potential for piping of the low plasticity, erodible core material into and through the unsealed near-surface joints resulting in failure of the dam was judged too dangerous even though the existing conditions were not considered nearly as bad as those causing failure at Teton Dam. It should be noted that no indications of uncontrolled seepage downstream or signs of embankment distress had been observed. The problem then became how to treat the various negative aspects of the existing situation to prevent the hypothesized failure.

DESIGN OF MODIFICATIONS

Foundation Drains. - The consultant's evaluation included suggestions on how to attack the potential erosion of core material out through near-surface joints in the foundation. Interception and drainage of these joints by filtered drains in the foundation were considered necessary to alter the existing hydraulic conditions. The greatest potential for damaging erosion of the core exists downstream of the grout curtain where no additional lines of defense were constructed. The best position for the foundation drains was therefore just downstream of the grout curtain, but far enough downstream of this barrier to guard against the high hydraulic gradients which could develop. The drains must also be oriented to intercept as many different formation beds and joints as possible. The orientation of the vertical joints at 45° to the valley in both directions made interception of the joints fairly easy. The drains had to be inclined upwards, both to intercept as many different beds as possible and to drain the seepage away from the embankment-foundation contact, particularly in the abutments. The drains also had to filter the core materials to prevent the drains from becoming piping conduits. The purpose of the drainage was to protect the remainder of the core contact downstream of the grout curtain against seepage at the contact. Such protection would still leave a substantial contact core length with little possibility of foundation seepage eroding zone 1 into the foundation joints. As a second line of defense, the extension of the existing toe drain up the abutments to the top of active conservation pool was judged the best solution. Seepage which might

travel as far as the downstream toe through near-surface abutment
joints should be intercepted by the toe drain.

The initial drainage ideas had to be developed into specifications
and had to be constructable. Drains could be drilled from the downstream abutment surfaces or from some access to the interior of each
abutment (tunnels or shafts). The length of the holes if drilled from
the downstream abutment surfaces created doubt as to correctly positioning the drain holes. Drain construction from the interior of the
abutments would shorten the lengths of the drilled drain holes. The
left abutment at Soldier Creek Dam contains two outlet works tunnels,
an upper and a lower (see fig. 1). These tunnels were about 130 feet
(40 m) apart vertically, and were 8-foot (2.4-m) diameter modified
horse-shoe tunnels. These tunnels provided access for drilling most
of the left abutment drain holes but were barely large enough for
drilling operations. A few drains had to be drilled from the upper
left abutment surface. Various systems of tunnels and shafts were
evaluated for right abutment access before selecting the one designed
and constructed (fig. 5). The inclined tunnel curves after developing
sufficient rock cover, then runs down the abutment at 18 percent slope
beneath the embankment to maintain an adequate depth of rock cover, and
is located parallel to and about 100 feet (30 m) downstream of the

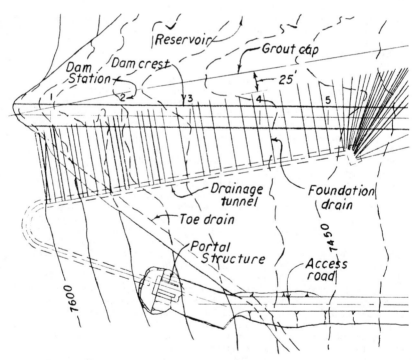

Figure 5. - Right Abutment Tunnel and Founation Drains

grout curtain. This tunnel is 11.5 feet (3.5 m) high and 12 feet
(3.7 m) wide for easier drain drilling and maintenance.
    The orientation of the drains to maximize interception of the
various formation beds and joints was the next feature designed. The
formation shales prevented vertical seepage communication. The tips
of the drains therefore had to be spaced close together vertically;
this spacing was chosen as 5 feet (1.5 m). The drain tips were also
positioned about 25 feet (8 m) downstream of the grout curtain. Each
formation bed is drained by a number of inclined drains in the design.
In order to ensure the drain intercepted all of the near-surface joints
possible, the drill holes were designed to hit the embankment core.
As-built payment sections at each 20-foot (6-m) station provided
excellent data on the location of the embankment-foundation contact.
Directional surveys of the alignment of each drain hole were required
to determine both horizontal and vertical position data on the actual
drain tip locations.
    The abutment drains were from 77 feet (23 m) to 460 feet (140 m) in
length which, together with the drain tip position accuracy specified,
required very accurate drilling. The directional surveys of the holes
determined whether or not additional drain holes were necessary for
proper abutment coverage. The specifications called for 4-1/4-inch
(108-mm) drain holes drilled using large diameter drill rods for stability and for inserting drainpipe through the rods if necessary. The
drill rods were 3-1/2-inch (89-mm) inside diameter by 4-inch (102-mm)
outside diameter flush joint casing. While insertion through the rods
was not needed on most drain holes, eight drains drilled and installed
from the upper left abutment surface required insertion of the drainpipe through the drill rods because of hole caving. Only 2 of
108 drain holes had to be redone because of incorrect hole alignment.
    The intercepted right abutment seepage had to be removed from a
chamber at the bottom of the tunnel. A 595-foot (181-m) drain hole
to provide gravity drainage from the chamber to the downstream toe of
the embankment was designed and was drilled from the toe in to the
chamber as a pilot hole which was then reamed to 14-inch (356-mm)
diameter. Minor caving of the reamed drill hole was experienced. A
10-inch (254-mm) diameter PVC sewer pipe with cemented flush joints
connecting the 5-foot (1.5-m) sections was inserted into the reamed
hole from the chamber and the pipe was fully grouted into the hole.
From an inspection well at the downstream embankment toe, a 12-inch
(305-mm) diameter gasket-joint PVC sewer pipe conducts the chamber
drain flow to the outlet works stilling basin. This chamber drain
system has a capacity of about 1,500 gal/min (6 m^3/min).
    The foundation drains in both abutments consisted of: (1) the
drilled 4-1/4-inch (108-mm) holes from the tunnels or abutment surface
to the core contact, (2) slotted 1-1/4-inch (32-mm) diameter schedule
80 PVC pipe, (3) Mirafi No. 140N nonwoven, polypropylene filter fabric
wrapped 2-1/2 times around the slotted PVC pipe, and (4) butyl rubber
"packers" located at certain positions along the pipe in the drain hole
(fig. 6). The pipe wrapped with filter fabric and the packers were
inserted into the drain holes to the core contact. The schedule
80 drain pipe with square threads could withstand 2,000 pounds (8.9 kN)
of force for pipe insertion purposes. The solid wall pipe annulus
adjacent to the tunnel wall or abutment surface was grouted.
    The drain hole alignment tolerances specified were +5 feet (1.5 m)
vertical and ±5 or 10 feet (1.5 or 3.0 m) horizontal, the latter

depending on the length of the drill hole. In addition, the drill
hole was not allowed to advance more than 10 feet (3 m) beyond the
specfied hole slope distance to prevent penetration through the grout
curtain.

The Mirafi filter fabric had a maximum fabric pore size (equivalent
opening size) of U.S. No. 70 sieve size. The design concluded that
this fabric pore size would properly filter all of the zone 1 and
zone 1 special compaction materials if eroded and carried to the filtered drainpipes. The filter fabric was attached to the drainpipe
using a hot-melt-type adhesive recommended by the fabric manufacturer.

Butyl rubber was used for the packers in order to resist potential
attack by acid in the seepage or groundwater intercepted by the drains.
The "packer" consisted of a hollow truncated cone with 1/4-inch
(6-mm) thick walls; the small cone end had a short sleeve portion which
slid over the 1-1/4-inch (32-mm) diameter, 1-foot (0.3-m) long PVC
pipe and was attached with a stainless steel clamp. Some packers used
one rubber cone and some used two cones. The packers isolated each
portion of the hole annulus along the pipe and centered the pipe in
the drill hole. The rubber cones were considered strong enough to prevent core material carried by seepage from bypassing the cone. The
two cone packers were used against the core at the end of the drill
hole, adjacent to the tunnel wall, and to isolate the annulus portions
in the zone of near surface rock joints where seepage was expected to
be the greatest. The cones were oriented with the small end toward
the embankment for ease of pipe insertion.

The abutment surface drains were completed with a "Y" fitting where
one leg connected to the toe drain and the other ran out to the surface

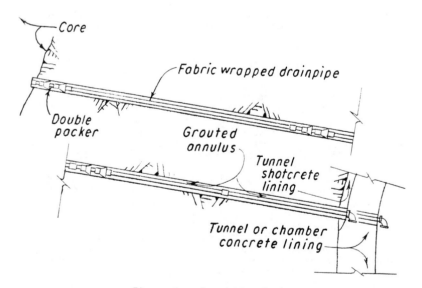

Figure 6. - Foundation Drains

for flow measurement. A valve in the toe drain leg is closed to direct
the drain flow out the other leg. The tunnel drains have a threaded
adapter on the ends to permit the attachment of a control system consisting of a valve and a pressure gage. If suspended material is found
in the flow from one or more drains, assumed to be piped core material,
the control system will be screwed onto the drainpipe adapter and the
valve will be closed to control the drain. The pressure gage would
read the water pressure at equilibrium to help determine the nature of
the problem. Remedial treatments would then be developed.
     Extended Toe Drains. - The extended toe drains design addressed two
problems. First, possible "tunnels" of segregated coarse material
along the the abutments in the zone 3 blanket drain had to be intercepted and filtered by the toe drain. Second, the toe drains had to
be keyed into the foundation rock where possible to provide the second
line of joint piping defense up to the maximum reservoir elevation.
     Interception of the existing blanket drain was accomplished by exca-
vating along the embankment toe to expose the blanket drain material
which was red in color. The perforated extended toe drain pipe was
surrounded by a two-stage filter with gravel around the pipe and sand
outside the gravel. On the lower right abutment, up to 35 feet (11 m)
of colluvium prevented keying the toe drain into rock. The upper ends
of the existing toe drains had to be exposed and the pipe had to be
continued. Inspection wells were constructed between the existing and
the extended pipes and at the upper end of the extended pipes. Ladders
and locking covers were incorporated in the inspection wells.

PERFORMANCE MONITORING

     The instrumentation, toe drains, and abutment drains are used to
monitor the dam's performance. Foundation and embankment piezometers
were added to the original instrumentation to determine surface surface conditions. The foundation drains appear to provide the best
data on foundation seepage conditions because the horizontal communication created by the bedrock joints permits each drain to measure its
seepage zone performance. Communication between seepage areas was
observed during drain drilling when flow from adjacent drain holes
would decrease when the same joint system was penetrated producing
flow out the drain hole.
     The seepage zones encountered correlated with the more jointed
sandstone beds but not necessarily the high grout take areas which
together had formed the basis for the drainage design. The right abutment tunnel construction encountered the largest portion of its inflow
in the same sandstone bed area where the drains were later most effective, totaling 65 to 75 gal/min (0.26 to 0.30 m^3/min). When the right
abutment drains were completed, the total flow from the drains was
90 gal/min (0.36 m^3/min) and tunnel was about 40 gal/min (0.16 m^3/min)
with the reservoir 160 feet (49 m) above the chamber. The left abutment drains, constructed first, totaled 170 gal/min (0.68 m^3/min) with
the reservoir water surface up to 166 feet (51 m) above the lower
tunnel. Reservoir fluctuations in 1983 and 1984 were reflected by
the left abutment drain flows, especially the drains with the largest
flows (fig. 7) which indicates the seepage comes from the reservoir.
Single drain flows as large as 37 gal/min (0.15 m^3/min) have been
recorded.

The design of the foundation drainage assumed that the hydraulic conditions downstream of the grout curtain would be changed and that the seepage arriving at the toe drains would be decreased. Piezometer water levels after, compared with before, the foundation drains were installed did not appreciably change. The toe drains flow had increased somewhat with reservoir rise to a winter level of about 40 gal/min (1.6 m³/min) but the construction of foundation drains, especially those in the right abutment, has significantly decreased the flow to a winter level of about 5 gal/min (0.02 m³/min). If the seepage intercepted by the left and right abutment drains has always passed through the foundation and out or into the blanket drain, it may be that a significant amount of seepage was bypassing the toe drains, either staying in the abutments or possibly passing under the toe drains. Since the foundation drains shorten the distance from the reservoir to a seepage exit and thus increase the hydraulic gradient for foundation seepage, an increase in the seepage flow has resulted.

SUMMARY AND CONCLUSIONS

The Soldier Creek Dam evaluations concluded that the potential for piping of erodible core material from the unsealed embankment-foundation contact out through foundation joints to the toe of the dam was low, but not negligible. Modifications were considered necessary to alter the hydraulic conditions in the foundation and to protect as much of the core contact downstream of the grout curtain as possible. A system of foundation drains was designed and constructed to intercept and filter foundation seepage as the best possible method of protecting the embankment core integrity. The existing toe drains were extended up the abutments to provide a second line of defense. Access to the

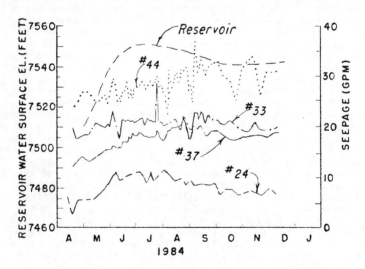

Figure 7. - Selected Left Abutment Drain Flows

interior of the abutments was available for most of the left abutment
drainage from the existing outlet works tunnels but had to be constructed in the right abutment with an inclined tunnel. The drain
holes had to be drilled with great care and precision to ensure complete coverage of the foundation. The engineering geology of a damsite must be well understood when designing foundation treatment
measures, especially those involving seepage defense.
    The foundation drains appear to be doing their work as a signific-
ant amount of seepage is intercepted by the drains and is safely
removed. The water flowing from the drains remains free of suspended
particles indicating that the filtering of the water entering the
drains is being done properly by the geotextile wrapped around the
drainpipes. The phreatic surface in the foundation remains below that
in the embankment which is desirable. The design of the modifications
correctly anticipated the materials and the construction techniques
required to properly complete the work. Present and future monitoring
adequately assesses the performance of the embankment and foundation
in light of the concerns for the dam's integrity. Soldier Creek Dam
is considered much safer as a result of the modifications constructed.
The total cost for the analysis, design, construction, and inspection
was $3.4 million.

ACKNOWLEDGMENTS

    The writer would like to thank Ralph B. Peck for his invaluable
work on the evaluation of and suggested modifications to Soldier Creek
Dam which shortened the process considerably. The writer would also
like to thank the U.S. Bureau of Reclamation which provided assistance
in the production of this paper and for whom the writer worked on this
project.

APPENDIX I. - REFERENCES

1. Acciardi, Raymond G., "Quantification of Pinhole Test Equipment
   Hydraulic Characteristics," U.S. Bureau of Reclamation,
   REC-ERC-82-15, September 1982.
2. Chadwick, Wallace L., et. al., "Report to U.S. Department of the
   Interior and State of Idaho on Failure of Teton Dam," Independent Panel to Review Cause of Teton Dam Failure, December 1976.
3. Peck, Ralph B., "Review of Soldier Creek Dam - Bonneville Unit -
   Central Utah Project," U.S. Water and Power Resources Service
   (USBR) Contract No. 1-07-DV-00140, April 13, 1981.
4. Peck, Ralph B., "Report on Proposed Remedial Measures, Soldier
   Creek Dam - Bonneville Unit - Central Utah Project," U.S. Bureau
   of Reclamation, Contract No. 2-07-DV-00153, January 25, 1982.
5. Wahler, W. A., and Associates, "Review of Design, Construction, and
   Operation of Soldier Creek Dam Project, Utah," U.S. Department
   of the Interior, Contract No. 14-01-0001-77-C-10, June 1977.

THE USE AND PERFORMANCE OF SEEPAGE REDUCTION MEASURES

By Robert D. Powell* and Norbert R. Morgenstern**,
    M.ASCE

Abstract

    The utilization and performance of various foundation
seepage reduction measures from over 100 dams situated on
pervious soils throughout the world has been reviewed.
Summary discussions of the utilization of slurry trench
cutoffs, concrete diaphragm walls, upstream impervious
blankets and grout curtains are presented. Also presented
is a compilation of acceptable and unacceptable seepage
quantities beneath dams, situated on previous soil
foundations, as represented in the literature. A criterion
for permissible seepage emerges that is of value for
preliminary design and in safety evaluation of embankment
dams.

Introduction

    The past utilization and performance of various
foundation seepage reduction measures beneath earth dams
founded on deep pervious foundations are valuable
precedents in design and performance evaluation. This is
especially true when selecting the most appropriate seepage
reduction measure in the preliminary stages of the design of
a new dam. In addition, it is of interest to explore
whether any general criteria with respect to seepage
volumes can be established as a guide to the review of
safety of such structures. Safety analyses on a periodic
basis are undertaken increasingly by owners of dams and by
regulatory authorities.

    In this paper the findings from the compilation and
correlation of published performance data on various
measures used to reduce and/or control seepage from over
100 dams throughout the world situated on pervious soil
foundations are presented. The interested reader is
referred to Powell (5), for specific information with
regards to pertinent references and details.

    The study concentrates on the applicability and
performance record of slurry trench cutoffs, concrete

*   Project Engineer, Gartner Lee Associates Limited,
    Buttonville Airport, Markham, Ontario, Canada, L3P 3J9
** Prof., Dept. of Civ. Engrg., Univ. of Alberta, Alberta
    Canada, T6G 2G7

diaphragm walls, upstream impervious blankets and grout curtains. Also included is a compilation of acceptable and unacceptable seepage quantities beneath dams on pervious soil foundations from published case histories throughout the world. It is not the intention of this paper to review at length various construction methods, specifications or stability considerations.

Seepage Reduction Measures

Dams were grouped according to the particular foundation seepage reduction measure used, viz., slurry trench cutoffs, concrete diaphragm walls including intersecting pile walls, panel walls and overlapped pile walls, upstream impervious blankets and grout curtains. These divisions were established so that performance and applicability of each measure could be assessed individually. However, the same data were collected for each measure with respect to the geometry of the dam, foundation conditions, construction details and performance, so that comparisons could be drawn.

Summary tables of data from the various case histories for slurry trenches, intersecting pile walls, panel walls, overlapped pile walls, upstream blankets and grout curtains are presented in Tables 1 to 6, respectively. The notation used in the tables are described in Appendix II.

In the course of assembling these data it became apparent that there were certain consistencies in the values of the depth of foundation, permeabiity of the foundation material and nominal hydraulic gradient for the various seepage reduction measures.

This was not unexpected, as these parameters are analogous to those used in Darcy's Law:

$$Q = kiA \dots\dots\dots\dots\dots\dots\dots (1)$$

in which, for the purpose of this study; Q = quantity of seepage; k = coefficient of permeability of the foundation soil; i = nominal hydraulic gradient defined as the difference between the forebay and tailrace level divided by the base length of the dam and A = cross-sectional area normal to the direction of flow or the depth per unit length of the foundation material to an impervious base. This analogy is shown on Figure 1.

Since the quantity of seepage beneath a dam was considered to be indicative of performance and the quantity of flow through the foundation soil is governed by Darcy's Law, the range of data for each seepage reduction measure were plotted with respect to permeability, nominal hydraulic gradient and depth of foundation material or area. See Figures 2, 3 and 4.

FIGURE 1: SCHEMATIC REPRESENTATION OF AREA,
NOMINAL HYDRAULIC GRADIENT AND
PERMEABILITY

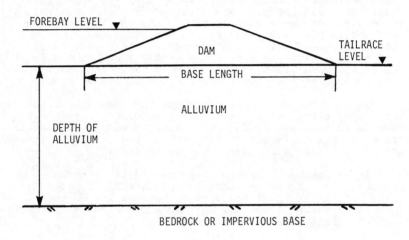

$$\text{NOMINAL HYDRAULIC GRADIENT} = \frac{\text{FOREBAY} - \text{TAILRACE LEVEL}}{\text{BASE LENGTH}}$$

AREA = DEPTH OF ALLUVIUM

PERMEABILITY = PERMEABILITY OF ALLUVIUM

TABLE 1

SLURRY TRENCH CASE HISTORIES - SUMMARY*

	DAM			SUBSURFACE CONDITIONS		S.R.M. DATA			LOCATION OF MEASURE	PERFORMANCE	
	DATE	HEAD (m)	NOMINAL GRADIENT	DEPTH (m)	k (m/s)	DEPTH (m)	WIDTH (m)	k (m/s)		EFFICIENCY (%)	ACCEPTABLE
Range											
Low	1952	4.9	.07	7	1×10^{-7}	4.5	1.3	1×10^{-9}	7 CL	60	1 No
High	1980	72.4	.26	350	1×10^{-2}	31	3.3	1×10^{-6}	9 u/s	89	8 Yes
Number	17	17	11	12	12	17	17	5	16	4	9
Average (Normal Dist.)	1967	22.2	.12	82	1×10^{-3}	18.7	2.4	2×10^{-7}	-	73h	-
Average (Lognormal Dist.)					6×10^{-5}			1.5×10^{-8}			

TOTAL NUMBER OF DAMS

17

* See notation used in Table within Appendix II.

TABLE 2
INTERSECTING PILE WALL CASE HISTORIES – SUMMARY*

	DAM			SUBSURFACE CONDITIONS		S.R.M. DATA			LOCATION OF MEASURE	PERFORMANCE	
	DATE	HEAD (m)	NOMINAL GRADIENT	DEPTH (m)	k (m/s)	DEPTH (m)	WIDTH (m)	k (m/s)		EFFICIENCY (%)	ACCEPTABLE
Range											
Low	1955	12	.08	10	$1{\times}10^{-6}$	22	.5	$1{\times}10^{-7}$	3 CL	60	–
High	1971	30	.27	40	$1{\times}10^{-2}$	41	.9	–	3 u/s	–	5 Yes
Number	6	6	6	6	3	6	6	1	6	1	5
Average (Normal Dist)	1962	20	.17	27.2	$3{\times}10^{-3}$	31	.65	$1{\times}10^{-7}$		60 q	
Average (Lognormal Dist)					$3{\times}10^{-5}$			$1{\times}10^{-7}$			

TOTAL NUMBER OF DAMS   6

* See notation used in Table within Appendix II.

TABLE 3

PANEL WALL CASE HISTORIES – SUMMARY*

	DAM			SUBSURFACE CONDITIONS		S.R.M. DATA				PERFORMANCE	
	DATE	HEAD (m)	NOMINAL GRADIENT	DEPTH (m)	k (m/s)	DEPTH (m)	WIDTH (m)	k (m/s)	LOCATION OF MEASURE	EFFICIENCY (%)	ACCEP-TABLE
Range											
Low	1959	20	.12	5	1×10^{-6}	18	.4	1×10^{-8}	6 u/s	60	2 No
High	1980	92	.27	140	1×10^{-2}	77	.76	1×10^{-8}	8 CL	100	8 Yes
Number	14	13	13	14	10	14	12	2	14	6	10
Average (Normal Dist)	1968	38	.19	53	1.1×10^{-3}	46	.63	1×10^{-8}		87q 85h	
Average (Lognormal Dist)					1.4×10^{-4}			1×10^{-8}			

TOTAL NUMBER OF DAMS

14

* See notation used in Table within Appendix II.

TABLE 4
OVERLAPPED PILE WALL CASE HISTORIES - SUMMARY*

	DAM			SUBSURFACE CONDITIONS		S.R.M. DATA			LOCATION OF MEASURE	PERFORMANCE	
	DATE	HEAD (m)	NOMINAL GRADIENT	DEPTH (m)	k (m/s)	DEPTH (m)	WIDTH (m)	k (m/s)		EFFICIENCY (%)	ACCEPTABLE
Range											
Low	1959	24	.12	27	1×10^{-5}	27	.5		2 u/s	82	2 No
High	1973	92	.34	122	1×10^{-2}	122	.66		4 CL	91	4 Yes
Number	6	6	5	6	4	6	6	0	6	2	6
Average (Normal Dist)	1965	54	.22	73	2×10^{-3}	67	.59	-	-	90q 87h	-
Average (Lognormal Dist)					1.6×10^{-4}						

TOTAL NUMBER OF DAMS   6

* See notation used in Table within Appendix II.

TABLE 5

UPSTREAM BLANKET CASE HISTORIES – SUMMARY*

	DAM			SUBSURFACE CONDITIONS		S.R.M. DATA						PERFORMANCE	
	DATE	HEAD (m)	NOMINAL GRADIENT	DEPTH (m)	k (m/s)	THICKNESS (m) min	max	X_t (m)	LENGTH (m) L	X_L	k (m/s)	EFFICIENCY (%)	ACCEPTABLE
Range													
Low	1937	12	.03	4.6	1×10^{-7}	.5	1	3.3	76	.6	1×10^{-9}	50	9 Yes
High	1978	138	.12	210	1×10^{-3}	3	12	92	1432	32	1×10^{-5}	65	10 No
Number	21	21	17	19	12	10	15	16	18	18	3	4	19
Average (Normal Dist)	1961	36	.06	61	2.4×10^{-4}	1.5	3.8	19	349	12	3×10^{-6}	58q 56h	–
Average (Lognormal Dist)					3×10^{-5}						1×10^{-7}		

TOTAL NUMBER OF DAMS

21

* See notation used in Table within Appendix II.

TABLE 6

GROUT CURTAIN CASE HISTORIES - SUMMARY*

	DAM			SUBSURFACE CONDITIONS		S.R.M. DATA				PERFORMANCE	
	DATE	HEAD (m)	NOMINAL GRADIENT	DEPTH (m)	k (m/s)	DEPTH (m)	WIDTH (m)	k (m/s)	LOCATION OF MEASURE	EFFICIENCY (%)	ACCEP-TABLE
Range											
Low	1953	7	.07	5	1×10^{-6}	7	2.4	1×10^{-7}	8 u/s	59	6 No
High	1977	122.5	.28	225	1×10^{-2}	225	38	1×10^{-4}	10 CL	98	12 Yes
Number	18	18	17	18	14	17	15	12	18	6	18
Average (Normal Dist)	1965	49	.16	69	1.3×10^{-3}	75	15	8×10^{-6}	-	80q 82h	-
Average (Lognormal Dist)					1×10^{-4}			7×10^{-7}			

TOTAL NUMBER OF DAMS

18

* See notation used in Table within Appendix II.

FIGURE 2: RANGE OF AREAS

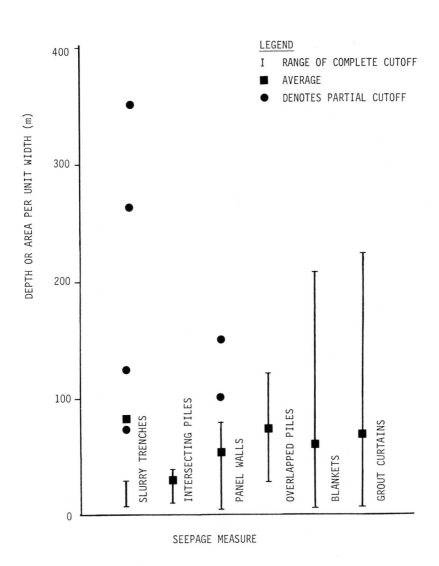

FIGURE 3:　RANGE OF NOMINAL GRADIENTS

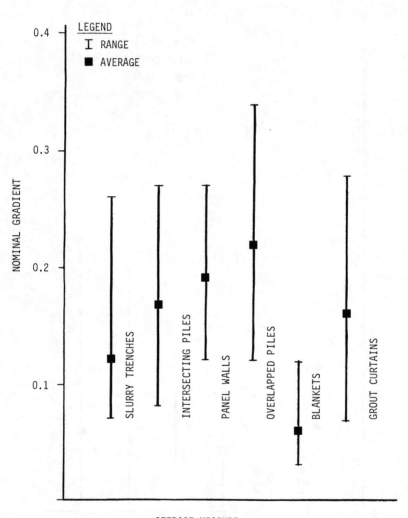

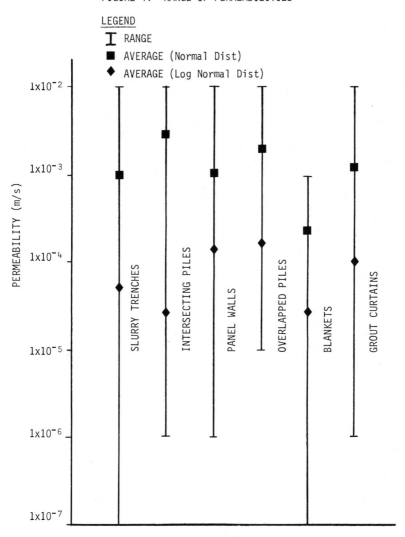

FIGURE 4: RANGE OF PERMEABILITIES

Slurry Trench Cutoffs

     A slurry trench cutoff as defined by Jones (4) and
for the purposes of this study, is a cutoff trench
excavated in the wet, supported using a bentonite slurry
and later backfilled with a blended soil while the slurry
is still in the trench. Slurry trenches have generally
been used as the shallowest form of complete cutoff and
have been constructed in a variety of foundation materials.

     The broad range of nominal gradients shown on Figure
3 would suggest slurry trenches may be constructed beneath
a wide range of dam profiles. However, as apparent in
Table 1, these dams are generally of low head and have
broad base lengths compared to those of other measures,
i.e., the average nominal gradient is 0.12.

     Of the 17 case histories reviewed, 56% located the
cutoff trench upstream, whereas 44% of the dams had the
trench located beneath the centreline. In all cases where
the information was available, the top of the trench was
flared in order to minimize the effects of differential
settlements between the granular alluvium and plastic soil
- bentonite mixture. Based on available data from the case
histories differential settlement has not been a problem.

     In approximately 60% of the case histories studied,
the cutoff trench was excavated and keyed into bedrock.
However, 40% of the walls were terminated in what was
considered to be an impervious stratum. The latter
trenches were designated as partial cutoffs and thus
thought to be less efficient. However, from available data
these walls were found to be just as effective in reducing
seepage as were the complete cutoffs.

     No trends could be established to suggest whether the
efficiency of slurry trenches is influenced by variations
in area, nominal gradient or permeabilty. Inefficiencies
were found generally to be related to lack of construction
control.

Concrete Diaphragm Walls

     For purposes of this study concrete diaphragm walls
have been separated into the following three categories:

(i)   Concrete Intersecting Pile Walls,
(ii)   Concrete Panel Walls, and
(iii) Concrete Overlapped Pile Walls.

     The common factor to each is that the excavation or
borehole is supported with a bentonite slurry during
construction and later backfilled with reinforcement and
a cement-bentonite mixture. The main differences between
each wall are their construction techniques and subsequent

plan views. Figure 5, demonstrates the typical plan views
of the three types of concrete diaphragm walls.

Intersecting Pile Walls

Intersecting pile walls were the fore-runners to
concrete diaphragm walls. They have been used under a
variety of dam profiles, as indicated by the range of
gradients on Figure 3. However, as in the case of slurry
trench cutoffs they have only been incorporated beneath
relatively low head dams, see Table 2.

Based on available data, intersecting pile walls have
been used when the foundation material is relatively
pervious and generally over a very narrow depth range due
to construction limitations.

With regard to the location of this seepage reduction
measure, 50% of the case histories were located upstream
and 50% were located beneath the centreline of the dam. In
all cases some form of plastic capping material was
placed at the top of the wall to reduce the risk of
cracking the core material.

All of the intersecting pile walls reviewed were
complete cutoffs to bedrock and were found to perform well.
However, due to construction limitations such as possible
misalignment with depth, the efficiency of this measure is
low in comparison with other diaphragm walls.

Panel Walls

Panel walls have normally been used for foundations
which were too deep to permit the use of intersecting pile
walls and too shallow to permit the use of overlapped pile
walls. As represented on Figure 3, by the narrower span of
gradients, the range of dam profiles under which panel
walls have been placed has been very restrictive in
comparison to most other seepage reduction measures.

Panel walls have been successfully constructed in
alluvium, containing 1 m diameter boulders, to depths
greater than 65 m.

Of the 14 case histories reviewed, and summarized
in Table 3, 60% of the panel walls were constructed beneath
the centreline of the dam, whereas 40% of the walls were
located upstream. The position of the measure does not
seem to have any influence on its performance record. As
in the case of intersecting pile walls, some form of
capping measure was constructed at the top of each panel
wall so as to protect the core material.

Panel walls have generally been used as complete cutoffs.
However, approximately 12% of the case histories were

FIGURE 5:   COMPARATIVE PLAN VIEWS OF CONCRETE
            DIAPHRAGM WALLS

INTERSECTING PILE WALL

PANEL WALL

OVERLAPPED PILE WALL

PLAN VIEWS

designated as partial cutoffs. The performance of these walls was also found to be good.

The efficiency of panel walls was found to be quite high and very consistent throughout the various settings in which they have been constructed. However, of the case histories studied, the performance of 20% of the walls was considered unacceptable. This is believed to be a reflection of design philosophy, in that 43% of the case histories relied solely on the effectiveness of the panel wall to reduce and control seepage quantities. Whereas with the other seepage measures studied, some additional form of seepage control measure such as relief wells was used.

Overlapped Pile Walls

As indicated on Figures 2 through 4, inclusive, overlapped pile walls have been used over a more defined range of foundation permeabilities, depths and dam profiles.

Generally they have been used when foundation depths exceed 55 m and the dam profile results in high nominal hydraulic gradients. Overlapped pile walls have also been successfully constructed through alluvium containing a high percentage of boulders.

Based on available data, 67% of the case histories located the cutoff beneath the centreline of the dam, whereas 33% of the cutoffs were situated upstream. Some form of pile cap had been provided for each of the case histories reviewed. The position of the cutoff did not tend to influence the effectiveness of this reduction measure.

Efficiencies, as indicated in Table 4, were found to be very high and consistent throughout the range of case histories. However, problems with long term performance were observed in 2 out of the 6 case histories studied, where seepage rates increased due to the deterioration of the overlapped pile walls.

Upstream Impervious Blankets

Upstream Blankets have tended to be used when the permeability of the foundation material is less than that of the other five seepage reduction measures discussed.

The average nominal hydraulic gradient, defined as the difference between the forebay and tailrace level divided by the base length of the dam including the length of upstream blanket, was found to be 0.06 or approximately 1/15.

Upstream blankets generally have had a poor track record, as indicated in Table 5. In over 50% of the case histories studied, the performance of upstream blankets has been unacceptable. Based on available data, 60% of the incidents took place at first filling and 40% at some later date.

Unacceptable behaviour was attributed to the following factors:

1)   Inadequate stripping of pervious foundation material.

2)   The blanket was not long enough as in the case of the Camanche 2 Dam as reported by Anton and Dayton (1).

3)   Inadequate seepage control measures had been installed to reduce uplift pressures at the toe of the dam.

In approximately 30% of the case histories reviewed, the quantity of seepage was found to decrease with time due to siltation behind the dam. However, it appears that if this decrease does not take place in the first three years after first filling, it will not occur.

Grout Curtains

    As shown on Figures 2 through 4 inclusive and in Table 6, grout curtains have been used under a wide variety of circumstances. However, on average they are used when foundation depths are greater than 70 m and the nominal hydraulic gradient is in the order of 0.16.

    Of the 18 case histories studied, 66% of the grout curtains were located beneath the centreline of the dam and 44% were located upstream. In all cases the grout curtains were complete cutoffs to bedrock. The position of the cutoff, also did not tend to influence its overall performance.

    The performance record of a grout curtain is similar to that of an overlapped pile wall, in that 67% of the case histories were effective and 33% were not. In two cases the efficiency of the grout curtains was also found to decrease with time. This was believed to be the result of progressive deterioration of the curtain. However, in three instances the quantity of seepage beneath the dam was found to decrease with time suggesting some form of healing or improved efficiency with time.

    In three of the case histories reviewed, the core of the dam above the cutoff cracked. However, this phenomenon was reported by Beier et al., (2) in the case of the Sylvenstein Dam to be the result of arching between the

steep abutments rather than due to the presence of the
grout curtain.

It is believed overlapped pile walls could have been
used in place of grout curtains in many instances.
However, choice is often a matter of personal preference
and is influenced by the availability of skills and
equipment.

Acceptable and Unacceptable Seepage

Many factors will influence whether or not a certain
magnitude of seepage is regarded as acceptable or
unacceptable. The allowable seepage will depend upon the
degree of localization and these values will be influenced
by the type of structure, the use of the reservoir, the
extent of habitation downstream and the owners. The
majority of dams reviewed here were part of hydroelectric
schemes and hence the economic value of the water from
these reservoirs would be comparable. The definition of
acceptable and unacceptable quantities of seepage beneath a
dam is very difficult to address. However, for the purpose
of this survey the distinction was based on the opinions of
various authors as reported in the literature.

The actual quantity of seepage beneath the dam was
recorded in two manners. Firstly, the total quantity of
seepage, Q, was tabulated for each dam. However, as one
would expect the quantity of seepage beneath a long dam
would generally be much greater than for a short dam and
therefore the seepage per unit length of structure was
calculated.

The acceptable and unacceptable values of seepage, as
described above, are plotted separately in histograms
presented on Figures 6 and 7.

Due to the range in values, the logarithm of the
quantity of seepage was plotted for convenience. The
median and mean values of the logarithms were found to be
essentially the same. However, the median value was
selected to be most representative of the data. The median
values of acceptable and unacceptable seepage are shown in
Table 7.

Undoubtedly many more cases exist for low and
acceptable seepage than reviewed in this study. Therefore
the histogram for acceptable seepage is not particularly
meaningful. However, the median value for the unacceptable
cases does indicate a level at which concern is raised.
From the available data this value appears to be 0.06 m^3/s
or 1.6 x 10^{-4} $m^3/s/ln$ m of dam. It is of interest to note
that in Norwegian practice Hoff (3) states that it is
considered unacceptable when the quantity of seepage
beneath dam is greater than 0.1 cubic metre per second. At

FIGURE 6:   ACCEPTABLE & UNACCEPTABLE SEEPAGE QUANTITIES, Q

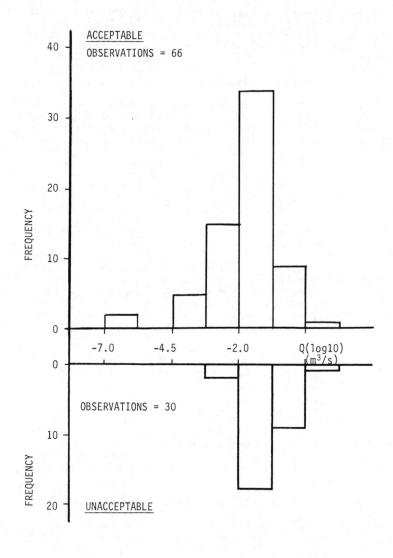

FIGURE 7: ACCEPTABLE & UNACCEPTABLE
SEEPAGE PER LINEAL METRE, Q/ln m

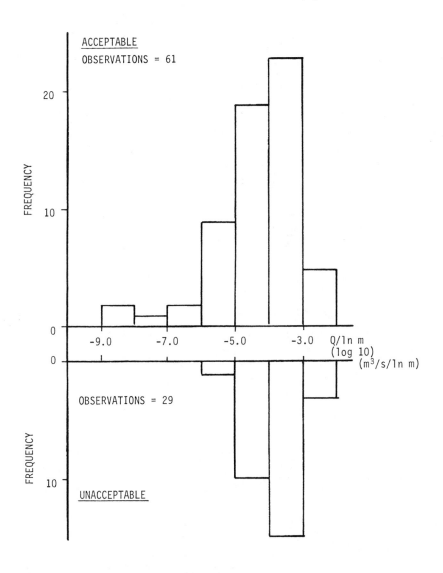

TABLE 7

ACCEPTABLE AND UNACCEPTABLE SEEPAGE

MEDIAN VALUES

	ACCEPTABLE	UNACCEPTABLE	
Q (m³/s)	0.029	0.064	Based on Available Data
Q/ 1n m	7.0×10^{-5}	1.6×10^{-4}	Based on Available Data
(m³/s / 1n m)			

that time he recommends that inspection of the dam be
increased to every other week rather than twice a year.

Another interesting factor which arose from a review
of these data is the time, in years after construction, at
which dams are subject to some form of increased seepage.
The distribution of the time of various incidents
after the completion date of the dam is shown on Figure 8.
The median time to unacceptable performance was 4.0 years.
As expected the majority of incidents take place at first
filling or in the following three years. However, it is
important to note the distinct presence of unacceptable
performance 50 years after construction.

Conclusions and Recommendations

This paper has examined in summary form the
performance of seepage reduction measures beneath earth
and rockfill dams on pervious soil foundations. The
examination consisted of a compilation and analysis of
available case histories throughout the world.

The following are the conclusions of this study:

1) Unacceptable seepage beneath a dam associated with a
hydroelectric scheme founded on alluvium is greater
than 0.06 m^3/s or 1.6×10^{-4} $m^3/s/ln$ m of dam. The
degree of unacceptability is governed by the order of
magnitude above these values.

2) The median time to some form of increased seepage or
incident after construction is 4.0 years. However,
unacceptable behaviour may occur even after 50 years of
acceptable performance.

3) Slurry trench cutoffs have been used effectively
beneath low head dams to depths of 30 m.

4) Intersecting pile walls were not found to be
the most efficient seepage reduction measure. However,
they have a good performance record.

5) Panel walls were found to be an effective
seepage reduction measure to depths of 60 m.

6) Overlapped pile walls were found to be the most
efficient method of sealing alluvium to depths of 120
m. However, their efficiency with time was found
to be more questionable.

7) Upstream blankets were not found to be the best
alternative seepage reduction measure in light of the
present, more efficient seepage reduction measures

FIGURE 8:   TIME OF UNACCEPTABLE SEEPAGE
            IN YEARS AFTER CONSTRUCTION

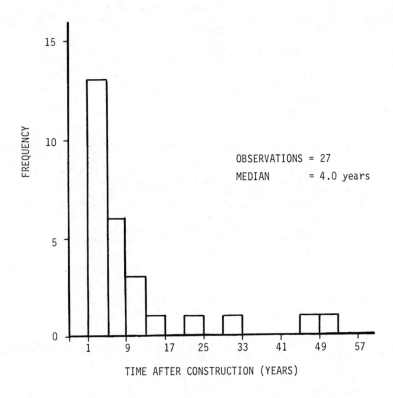

available. Additional efficiency because of
sedimentation behind the dam does not always occur.

8) Grout curtains were found to have a similar performance
record to that of overlapped pile walls. However,
the efficiency of grout curtains was approximately
10% less.

9) The optimum location of a cutoff wall is not governed
by the efficiency of seepage control.

Acknowledgements

Thanks are expressed to Messrs. A.G. Hims, P.K. Lee,
M.ASCE and the personnel of Gartner Lee Associates Limited
for their help in producing this paper. This research was
carried out with the financial assistance of the National
Research Council of Canada at the University of Alberta.

Appendix I - References

1) Anton, W.F. and Dayton, D.J. 1972. Camanche Dike-2
Slurry Trench Seepage Cut-off. American Society
of Civil Engineers Conference, Purdue, pp.
735-749.

2) Beier, H., List, F. and Lorenz, W. 1979. Subsequent
Sealing in the Impervious Core of the Sylvenstein
Dam., 13th International Congress on Large Dams,
Vol. 2, Question 49, Response 8, New Delhi, pp.
103-116.

3) Hoff, Th. 1970. Supervision of Dams and Reservoirs
in Operation Owned by the Norwegian Government.
10th International Congress on Large Dams, Vol. 3,
Question 38, Response 47, Montreal, pp. 881-898.

4) Jones, J.C. 1967. Deep Cut-offs in Pervious Alluvium
Combining Slurry Trenches and Grouting. Volume
I, 9th International Congress on Large Dams,
Question 32, Response 31, Istamboul, pp. 509-524.

5) Powell, R.D. 1984. "Performance of Seepage Measures
Beneath Earth and Rock Fill Dams on Pervious Soil
Foundations". Unpublished M.Sc. Thesis,
University of Alberta.

Appendix II - Notation

A   = Area in metres
CL   = Centreline location of cutoff
h   = efficiency based on head drop
i   = nominal hydraulic gradient
k   = coefficient of permeability in metres per second

m	=	metre
m/s	=	metre per second
m^3/s	=	cubic metre per second
$m^3/s/ln\ m$	=	cubic metre per second per lineal metre of dam
q	=	efficiency based on decreased seepage
Q	=	quantity of seepage in cubic metre per second
Q/1n m	=	quantity of seepage per lineal metre of dam
S.R.M.	=	Seepage Reduction Measure
u/s	=	upstream location of cutoff
date	-	The year when the dam was completed and ready for use
Head	-	Difference between the forebay and tailrace level
Nominal Gradient	-	The hydraulic gradient, is the Head divided by the base length of the dam
Subsurface Conditions	-	Information with respect to the foundation soils
Depth	-	The depth of soil beneath the dam to bedrock
k	-	The permeability of the foundation soil
S.R.M. Data	-	Seepage Reduction Measure Data
Depth	-	The depth to which the cutoff extends
Width	-	The width or diameter of the cutoff.
Location of Measure	-	The position of the cutoff beneath the dam. (CL - centreline, u/s upstream).
k	-	Permeability of the seepage reduction measure.
Thickness	-	Thickness of upstream blanket.
min	-	Minimum thickness of upstream blanket.
max	-	Maximum thickness of upstream blanket.
Xt	-	Thickness of the upstream blanket divided by the head.
Length (L)	-	Length of the upstream blanket from the upstream toe of the dam.
XL	-	Length of the upstream blanket divided by the head.
Efficiency & Performance	-	The observed percent efficiency and behaviour of the seepage reduction measure.
		q - efficiency based on quantity of seepage
		h - efficiency based on head loss

Designing and Monitoring for Seepage at Calamus Dam

By Errol L. McAlexander, M. ASCE[1] and William O. Engemoen[2]

ABSTRACT: Given the complex, pervious nature of the foundation for Calamus Dam, every effort has been made to incorporate design features that will reduce and control the anticipated large volumes of underseepage and high exit gradients. An extensive instrumentation system has been included in the design in order to monitor the effects of this seepage and the performance of the dam. As the reservoir fills and begins to operate, it is believed that the instrumentation will provvide valuable data to enable a better understanding of seepage behavior in this type of foundation, evidence of how well various design features function, and knowledge of how well design tools such as finite element seepage modeling can predict behavior.

INTRODUCTION

Calamus Dam and Reservoir offered the challenge of designing a dam on up to 150 feet (46 m) of erodible silts, sands, and fine gravels. The major problem presented by these variable alluvial materials was the control of potential large volumes of seepage and accompanying high exit gradients. Seepage analyses were performed to estimate the volumes of flow anticipated and the potentials anticipated throughout the embankment and foundation. Seepage control measures incorporated in the design of the dam included extensive foundation excavation, an impervious upstream blanket, a slurry trench cutoff wall, and a downstream drainage system which included a toe drain, relief wells, collector drains, and outlet channels and stilling basins lined with filter/drainage materials. Once the reservoir begins filling, the effectiveness of these design features to control seepage and ground water conditions will be monitored through the use of pneumatic, porous-tube, and vibrating-wire piezometers, observation wells, a thermal monitoring network, weirs, and turbidity metering units.

PHYSICAL DESCRIPTION OF CALAMUS DAM AND APPURTENANT WORKS

Calamus Dam is located in Garfield and Loup Counties (shown in figure 1) near the town of Ord in north central Nebraska. This Bureau of Reclamation project is part of the North Loup Division of the

---

[1]Supervisory Civil Engineer, U.S. Bureau of Reclamation, Engineering and Research Center, Embankment Dams Branch, Denver, Colorado 80225.

[2]Civil Engineer, Principal Designer, U.S. Bureau of Reclamation, Engineering and Research Center, Embankment Dams Branch, Denver, Colorado 80225.

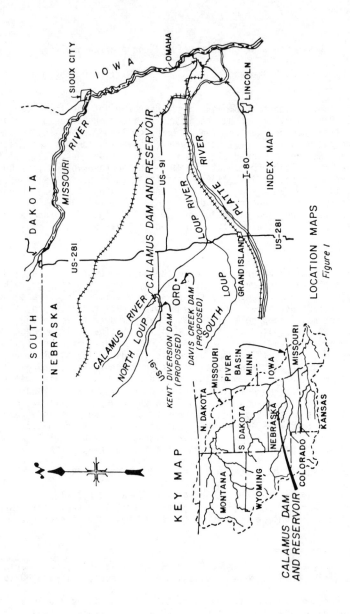

LOCATION MAPS
Figure 1

Pick-Sloan Missouri Basin Program. The project purpose is to develop
and manage water of the Calamus and North Loup Rivers for irrigation,
recreation, fish and wildlife enhancement, and flood control. The
North Loup Project includes Calamus and Davis Creek Dams, Kent
Diversion Dam, a pumping plant, five principal canals, and several
laterals.

Calamus Dam, shown in figures 2 and 3, has a structural height of
95 ft (29 m), a crest length of 7,200 feet (2,200 m), and a crest width
of 30 ft (9.1 m). Waterways include a spillway and a river outlet
works with respective capacities of 3,080 ft^3/s (87 m^3/s) and
2,460 ft^3/s (70 m^3/s) at a maximum water surface elevation of 2252.8 ft
(687 m). The reservoir storage at the top of active conservation,
elevation 2244.0 ft (684 m) is 100,000 acre-feet (123,000,000 m^3). A
surcharge pool of 50,200 acre-feet (62,000,000 m^3) in combination with
the spillway and river outlet discharges protects against the inflow
design flood which has a peak of 57,000 ft^3/s (1610 m^3/s) and a 3-day
volume of 69,600 acre-feet (86,000,000 m^3).

Calamus Dam is being constructed under three contracts. Stage 1
(a $1.1 million contract) was completed in 1981 and consisted of
1,200,000 yd^3 (917,000 m^3) of foundation excavation and selective
stockpiling for future use.

Stage 2 is under construction and consists of the major portion of
the main embankment; including grading and construction of an upstream
blanket, construction of a slurry trench, lining of an intake channel
with impervious embankment and soil-cement, excavation for and
construction of the river outlet works and spillway, construction of
filters around stilling basins, a filter/drainage layer and a crushed
rock armor layer in the outlet channels for the river outlet works and
spillway, excavation for and construction of a baffled apron drop,
foundation trench excavation, and embankment construction to full
height, including soil-cement slope protection. Industrial
Constructors Corporation of Missoula, Montana, with a low bid of
$35.4 million, is the prime contractor.

The Stage 3 contract is also under construction and includes 44
relief wells, a toe drain, the remainder of the embankment, a
downstream horizontal drainage blanket, and river channel fill from the
downstream toe of the dam to the confluence of the outlet channels.
The stage 3 contract was awarded for $4.4 million, with the Stage 2
contractor being the successful low bidder. Stage 2 and Stage 3 are
being constructed concurrently.

GEOLOGY

Regional Geology. - Calamus damsite is near the southeastern edge of
the 20,000 mi^2 (52,000 km^2) Sandhills region of Nebraska. The
Sandhills are a rolling expanse of sand dunes and interdune depressions
that overlie extensive alluvial and fluvial deposits which in turn
overlie the Ogallala Formation, an extensive aquifer in the central
United States consisting of weakly indurated sands and silts
interlayered with unconsolidated silts, sands, and gravels.

Damsite Geology. - The Calamus River Valley consists of the river
channel, the flood plain, and low terraces that extend toward the sand

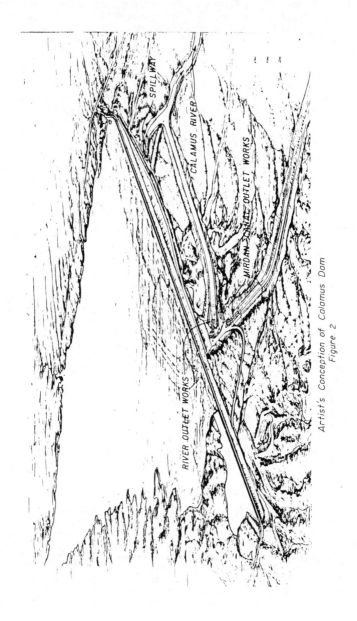

Artist's Conception of Calamus Dam
Figure 2

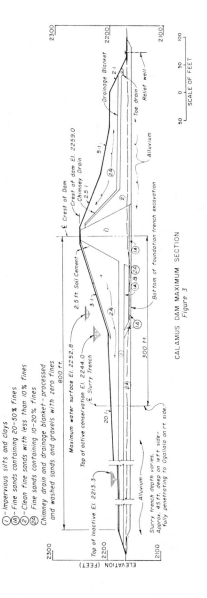

CALAMUS DAM MAXIMUM SECTION
Figure 3

dune ridges that form the abutments. The surficial materials overlying the Ogallala Formation consist of fluvial deposits, including peat deposits in buried oxbows. In the area of the present-day Calamus River, these older fluvial deposits have been incised by the river and more recent alluvial-fluvial deposits have been redeposited in the river channel. On the abutments, the alluvial-fluvial deposits have been capped by sand dunes that were derived from wind erosion of the extensive Ogallala Formation and alluvial and fluvial deposits in the area. Surficial deposits are 30 to 150 feet (9 to 46 m) thick and, as shown in figure 4, are subdivided into units 1 through 8 on the basis of geology and their corresponding engineering properties. The impervious Pierre shale occurs at a depth of approximately 400 feet (122 m).

Geologic Units. - Unit 1 is eolian in origin and is predominantly fine sand that mantles the valley except in the recent river channel. The dune sand consists of fine to medium sand with generally less than 5 percent silty fines. The greatest thickness of this deposit is approximately 45 feet (14 m) at the abutments.

Unit 2 loess consists of nonplastic silt with varying amounts of fine sand and some plastic clay strata. This unit generally underlies the sands of unit 1 on the left abutment. The maximum thickness is approximately 50 feet (15 m) in an interdune trough on the left abutment. The loess deposits vary in density between 82 and 104 lb/ft^3 (1310 and 1670 kg/m^3).

Unit 3 includes black, fibrous peat deposits, organic silty sands, silts, and lean clays. These deposits vary in thickness up to 15 ft (4.6 m), but more typically are up to 10 ft (3.0 m) thick. This entire sequence is compressible. This unit is found in a layer on each side of the river as shown in figure 4.

Unit 4 consists of the alluvial deposits of the Calamus River. The alluvium is predominantly interbedded fine sands and some silts of up to 130-ft (40-m) depth. The percentages vary significantly from stratum to stratum.

Unit 5 is a sequence of interbedded silty sands and uniformly graded sands of fluvial origin. Some medium sand and fine gravels are interbedded with fine and medium sands. This unit varies in thickness up to approximately 100 ft (30 m) on the left abutment and up to 20 ft (6 m) on the right abutment.

Unit 6 consists of fine sand to fine gravel generally containing less than 5 percent fines. The thickness varies up to 85 ft (26 m) on the left abutment and up to 45 ft (14 m) thick on the right abutment. Unit 6 forms the foundation for the river outlet works on the right abutment, and is the most pervious deposit in the foundation.

Unit 7 consists of fine sands and silts which overlie the Ogallala Formation. The thickness of this unit varies up to 20 ft (6 m).

Unit 8, the Ogallala Formation, is bedrock at the site and consists of interbedded, weak, variably indurated sandstones and siltstones. Unconsolidated fine sands and silts are interlayered with the variably indurated formation. A 5-ft (1.5-m) thick hard pan cap is at the top of the Ogallala formation.

The ground water conditions at the site are variable. Artesian pressures extending to within 5 ft (1.5 m) above ground level were

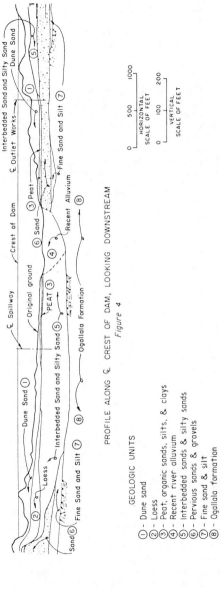

PROFILE ALONG ℄ CREST OF DAM, LOOKING DOWNSTREAM

Figure 4

GEOLOGIC UNITS

① - Dune sand
② - Loess
③ - Peat, organic sands, silts, & clays
④ - Recent river alluvium
⑤ - Interbedded sands & silty sands
⑥ - Pervious sands & gravels
⑦ - Fine sand & silt
⑧ - Ogallala formation

measured in the Ogallala Formation (and influenced dewatering for the
outlet works stilling basin located at the right abutment). Perched
water was trapped in unit 1 by the less pervious units 2 and 3 that
underlie unit 1. Units 2 and 3 also trapped water and were generally
wet. The ground water in the abutments was generally at about elevation 2200 ft (671 m). The ground water was fed by rainfall
infiltrating vertically to the ground water table.

EXPLORATIONS AND TESTING

During the Stage 1 construction, the partial excavation of the foundation trench, along with a 650-ft (198 m) long exploratory trench,
provided an evaluation of foundation materials and confirmation of
design assumptions. The exploratory trench excavation encountered predominantly fine to medium sand with less than 5 percent fines. One to
three ft (0.3 to 0.9 m) of peat overlayed the sands with the peat
overlain by a silty sand mantle. Data obtained from testing samples
taken from these trenches enabled the designers to confirm high
compressibility of the peat layer, the nature and permeability of the
alluvial-fluvial deposits, and strength assumptions used for stability.
In order to evaluate piping potential, numerous gradation, density,
and permeability tests were run on the various foundation materials.
Terzaghi filter criteria were applied to evaluate piping potential between adjacent foundation materials. In all cases evaluated, piping
criteria were satisfied between adjacent zones. Filter criteria were
also used for designing material interfaces within the embankment. In
addition, a worst case filter test was run in the laboratory to verify
filter criteria. Under gradients of 1, 4, and 16, no fines were lost
between zone 1A material (fine sand with more than 20-percent fines)
and coarse foundation sands.

FOUNDATION CONDITIONS

The foundation is a complex eolian-alluvial deposit. The dune
sands, alluvial sands, silts, clays, and peat were deposited, mixed,
and weired developing dunes, layers, lenses, and pockets of materials
of various proportions and properties. Figures 4 and 5 show the
general distribution of these materials.
The fibrous peat layer varied in extent as shown in figure 5. The
fibrous peat was found in layers up to 4.5-ft (1.4-m) thick. The test
trench and borings showed there was generally no continuity in the
thickness or extent. Ancient meanders may have cut through previous
deposits and peat deposits developed in oxbow remnants. Frequently
associated with the peat deposits were additional layers of peat-
contaminated materials above and below the peat and ranging up to 5-ft
(1.5-m) thick. The variability of the thickness and continuity of peat
and peat contaminated deposits was cause for considerable concern. The
peat was tested and obviously indicated extensive compression, along
with potentially low strength. The low strength, the variability in
compressibility and extent, and the concern for cracking due to differential settlements in combination with fine sand and loess construction materials resulted in the decision to remove the peat from beneath
the impervious core and from beneath extensive portions of the fine
sand shell materials.

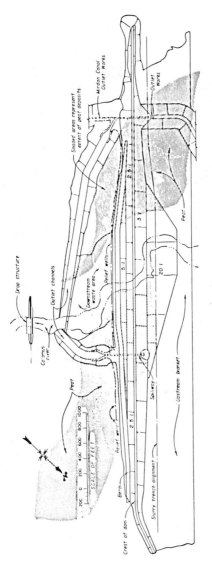

EXTENT OF PEAT DEPOSITS
Figure 5

A clay stratum on the left abutment that was a continuation of the peat layer was examined closely during preconstruction excavation of test pits. The layer was sampled and tested for physical properties. Following review of gradation analyses, the clay was tested for dispersive characteristics. Some material was found to be dispersive. The materials immediately overlying and underlying the clay were sampled and gradations were determined. Upon analyzing filter criteria, movement of clay materials through either of the encompassing materials was found to be unlikely. The foundation trench was extended to cut off the continuity of this clay layer. The treatment is expected to confine the dispersive materials and prevent their movement, minimize consolidation and strength concerns with this layer, and improve downstream drainage.

A loess deposit underlies the dune sand in a portion of the left abutment. Testing of undisturbed samples indicated these materials could be susceptible to collapse under a certain embankment loading upon saturation during first filling of the reservoir. The embankment will be less than 15 ft (3 m) high in this area, which should not be enough to produce significant settlement. However, in order to induce consolidation during construction, prewetting of these materials was specified. The results will be monitored with a settlement plate and surface measurement points.

SEEPAGE ANALYSES

A three-dimensional finite element analysis was performed to evaluate seepage conditions. The three-dimensional finite element computer program SEEP3D was used to calculate seepage quantities, establish potential seepage exit locations, and calculate gradients. The effects of the outlet channels for the outlet works and spillway on seepage and the corresponding effectiveness of design features such as the slurry trench, impervious upstream blanket, chimney drain, toe drain, relief wells, and downstream drainage blanket in controlling the seepage were modeled.

The finite element grid was divided into right and left bank models, with the Calamus River selected as a zero flow boundary. More than 2,600 elements and 3,400 nodes were used to model each side of the river (the right valley and left valley). All models were run for steady-state (equilibrium) conditions at reservoir elevation 2,244 ft (684 m), which is the top of active conservation pool. The models had four layers of elements. The lower layer of elements represented the Ogallala Formation and the top three layers the other geologic units listed in table 1.

Table 1

Foundation Properties Used in Finite Element Model

Geologic unit	Density (lb/ft^3)	Permeability K_H ft/yr	(cm/s)	K_H/K_V
1, 2, and 3	82 to 104	15,000	(0.014)	1.5
4	104 to 118	30,000	(0.029)	25
5	104 to 115	12,000	(0.012)	50
6	100 to 118	140,000	(0.135)	10
7 and 8	92 to 126	3,500	(0.003)	25

Reservoir potential was applied on the reservoir floor. The upstream blanket was modeled as an impermeable boundary where seepage pressures were allowed to build. Steady-state water table elevations were allowed to develop in the pervious zones of the embankment and in the downstream alluvium. The relatively impermeable top surface of the Pierre Shale was modeled at elevation 1,800 feet (549 m).

Full reservoir potential was conservatively imposed vertically on the upstream boundary of the model throughout the total thickness of the alluvium. Potentials determined from the results of a regional ground water model were used for the side and downstream boundary conditions. A narrow row of elements with a permeability of 1 ft/yr (1 x 10^{-6} cm/s) represented the slurry trench. The internal drainage blanket was modeled using a permeability of 140,000 ft/yr (0.14 cm/s). Zero potential at invert elevations 7 feet (2.1 m) below the average ground surface was used for the toe drain and relief wells. Hydraulic potentials equal to ground surface elevation were used for seep areas downstream from the dam.

The outlet works stilling basin and channel are the dominant drainage features at the site because they penetrate through the highly permeable, coarse sand of geologic unit 6. The spillway and outlet works underdrainage was modeled with an assigned potential of tailwater elevation 2,171 ft (662 m). Studies predicted at least 10 ft (3.0 m) of degradation below streambed. This was used in modeling the river channel downstream from the baffled apron drop structure.

Results obtained from the seepage analyses show the effect of the slurry trench and relief wells on the seepage flow (table 2). Case 1 establishes the seepage condition for no slurry trench or relief wells. The effect of the slurry trench was modeled in Case 2 (slurry trench) and Case 3 (slurry trench and relief wells). Table 2 gives the breakdown in quantities of seepage flows at various key locations for the three cases modeled and shows minimal effect of the slurry trench for reducing seepage quantity. This was in part due to the relatively high permeability of the Ogallala formation. The maximum calculated vertical gradients occurring in the foundations materials at key exit points are presented in table 3 and indicate potential for movement of soil grains if adequate filters are not required. Water table elevation contours with and without the slurry trench and relief wells are shown on figure 6.

The slurry trench is more effective on the right bank because the trench penetrates the Ogallala Formation and cuts off the highly permeable, gravelly sand of geologic unit 6. Substantial reduction of seepage flows by a slurry trench cutoff at the damsite was not feasible because of the high permeability (table 1) of the underlying Ogallala formation (unit 8) and the very high permeability of the coarse sand units which extend indefinitely beyond both abutments of the dam. The total estimated seepage loss at the dam is approximately 37 ft^3/s (1.05 m^3/s) without the slurry trench and relief wells. With the slurry trench completely penetrating the sand and gravel unit on the right side of the river channel, the total estimated flow is reduced to 32 ft^3/s (0.9 m^3/s). The partially penetrating slurry trench underneath the entire length of the left side of the dam in the interbedded fine sands reduces the estimated flow by only 0.3 ft^3/s. The fully penetrating slurry trench cutoff to Ogallala formation is effective in reducing calculated exit gradients in the right bank, but less effective for the partial penetrating slurry trench in the left bank.

Table 2

Predicted seepage flows in L/s (ft^3/s)

	Case 1		Case 2		Case 3	
	Rt bank	Lt bank	Rt bank	Lt bank	Rt bank	Lt bank
TO	725(25.6)	328(11.6)	593(20.6)	320(11.3)	592(20.9)	371(13.1)
UD	555(19.6)	190(6.7)	402(14.2)	178(6.3)	408(14.4)	227(8.0)
AD	170(6.0)	139(4.9)	184(6.5)	142(5.0)	184(6.5)	144(5.1)
SB	224(10.2)	20(0.7)	184(8.2)	20(0.7)	181(8.1)	11(0.4)
CH	280(9.9)	25(0.9)	227(8.0)	23(0.8)	204(7.2)	14(0.5)
TD	51(1.8)	62(2.2)	28(1.0)	57(2.0)	6(0.2)	25(0.9)
RW	-	-	-	-	57(2.0)	136(4.8)
SS	-	6(0.2)	-	6(0.2)	-	0(0.0)
RC	71(2.5)	96(3.4)	71(2.5)	96(3.4)	68(2.4)	79(2.8)
DB	156(5.5)	144(5.1)	150(5.3)	144(5.1)	144(5.3)	136(4.8)

Key for tables 2 and 3:

TO	= Total seepage flow.
UD	= flow under the dam.
AD	= flow around end of dam.
SB	= flow exiting into outlet works or spillway stilling basin.
CH	= flow exiting into outlet works or spillway channel.
TD	= flow exiting into toe drain.
RW	= flow exiting into relief wells.
SS	= flow exiting as surface seepage.
RC	= flow exiting into modeled portion of river channel.
DB	= flow across downstream model boundary.

Case 1 - No slurry trench, no relief wells.
Case 2 - Slurry trench, no relief wells.
Case 3 - Slurry trench and relief wells.

Table 3

Maximum exit gradients

	Case 1		Case 2		Case 3	
	Rt bank	Lt bank	Rt bank	Lt bank	Rt bank	Lt bank
CH	1.2	1.0	1.1	1.0	0.9	1.0
SB	0.7	2.1	0.4	2.0	0.4	1.4
TD	3.6	4.0	2.6	3.5	0.7	2.3
SS	-	0.0	-	0.0	-	0.0

DESIGN

Numerous design features were incorporated into Calamus Dam in order
to handle the anticipated seepage problems indicated by the geologic
investigations and seepage analyses. Most of the features are shown in
figure 3. The following are descriptions and purposes of the various
features.

Foundation Trench. - The foundation trench is an important design
feature located beneath the entire length of the embankment and
extending approximately 330 ft (100 m) into each of the sand dunes
located on both the right and the left abutments. The foundation
trench will cut off low density sand dune deposits in both abutments, a
soil stratum containing dispersive clay materials on the left abutment,
moderate and low density loess on the left abutment, lower density
silty sand and fine sands which mantle the river valley, and extensive
beds of highly compressible peat located in both banks of the river.
The foundation trench will minimize differential settlement and potential of embankment cracking, and will serve to cut off surficial pervious deposits, thus minimizing seepage and piping concerns.

The loess deposit on the left abutment is too deep to completely
remove economically. Consequently, the area was treated by prewetting.
However, it is believed that the height of embankment will be too low
to initiate any significant settlement of the loess even when
saturated. The loess is protected against piping by the natural peat
and clay blanket and the slurry trench.

Upstream Blanket. - Natural deposits of peat, loess, and clay offer
varying degrees of impervious blanketing over the reservoir floor.
Permeabilities of the natural blanket are highly variable and the
recent reworked river alluvium interrupts the natural blanket and
intersects the pervious sands on the right and left banks of the river.
It was thus considered necessary to include an impervious blanket,
generally extending 800 to 1,300 ft (240 to 400 m) upstream of the centerline crest of the dam for the full length of the dam, in order to
control entrance conditions for seepage flow into the foundation and
increase the length of the seepage path through the foundation.
Figures 3 and 5 show the extent of the upstream blanket.

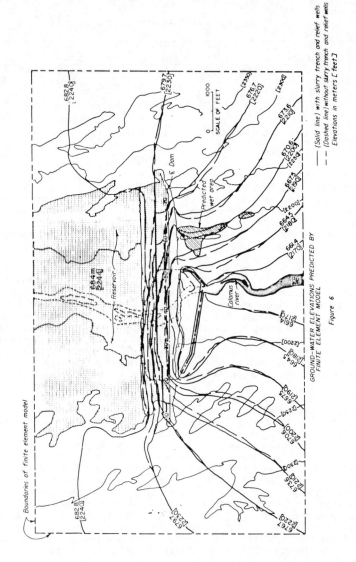

GROUND-WATER ELEVATIONS PREDICTED BY
FINITE ELEMENT MODEL

Figure 6

—— (Solid line) with slurry trench and relief wells
— — (Dashed line) without slurry trench and relief wells
      Elevations in meters [feet]

Slurry Trench Cutoff Wall. - The decision to construct a slurry trench cutoff wall was made following extensive analyses and a design review by a panel of consultants[3]. Preliminary model studies indicated an approximate reduction in total seepage of 10 percent for a fully penetrating slurry wall to the Ogallala Formation constructed for the entire length of the dam. One very positive effect the slurry trench not reflected in the model is that it will ensure that anomolies in natural flow channels (carrying potential con centrated seepage flows) will be cut off. The major advantage of the slurry wall is that it will provide a physical barrier in the alluvium and minimizes potential for uncontrolled seepage through the foundation.

The length of the slurry trench is over 7,000 feet (2,134 m), and stretches from the far right abutment to the far left abutment. The depth of the slurry wall on the right bank is approximately 105 ft (32 m) - the cutoff wall is fully penetrating on this side. The foundation on the left valley consists in part of natural silts, clays, and peat layers forming a semicontinuous natural blanket. These materials will be intersected by a partially penetrating slurry wall 45 ft (14 m) deep. The cost of the additional excavation for a fully penetrating slurry wall on the left valley was not cost effective. The combination of the fully penetrating slurry wall on the right bank with a partially penetrating slurry wall on the left bank (both located 300 ft (91 m) upstream of dam centerline) was finally selected as the most effective treatment.

Toe Drain. - The reinforced concrete pipe toe drain will intercept near surface seepage near the toe of the dam. The toe drain was designed using a safety factor of five times flows predicted by the finite element model studies.

Relief Wells. - Control of seepage pressures at the toe of the dam is a critical element of the design which is accomplished by the relief wells. These wells were designed for a maximum flow of 0.5 ft^3/s (0.01 m^3/s). The outflow elevations and spacings were established by using data from the finite element model in conjunction with convential relief well design practices. The alinement of the wells is shown in figure 5. Additions to the relief well system will be accomplished, as required, depending upon the results obtained from monitoring the wells and associated instrumentation.

Drainage Blanket and Chimney Drain. - The chimney drain and drainage blanket will control seepage within the embankment and underseepage near the downstream toe. However, the most important function is protection against piping in the event of embankment cracking. Traditional Terzaghi filter and piping criteria were used to establish the gradation for these drainage materials.

Appurtenant Works Structures. - Both the river outlet works and spillway are constructed through the embankment. Design concerns were piping along the conduit, deformations due to embankment loadings, and control of the seepage along the river outlet works alinement. The intake structure is underlain by 10 feet (3 m) of impervious material and a two-stage filter constructed in two 3-ft layers. A steel lining

---

[3]Panel of Consultants included: Thomas Leps, George Sowers, and Clifford Cortwright.

in the conduit upstream from the gate chamber of the river outlet works and in the entire spillway conduit was added as a precaution against piping of material into the conduit and introduction of high water pressures in the embankment should the conduit separate due to settlement. The linings will be welded following construction of the embankment, by which time most of the anticipated settlement and consolidation is expected to have occurred. A steel outlet pipe within the concrete access conduit downstream from the gate chamber will conduct flow to the downstream control house. The stilling basins are lined with a three-stage filter as an additional defensive measure to control anticipated high exit gradients and seepage flows into the basin areas.

INSTRUMENTATION FOR MONITORING SEEPAGE

The concern for seepage, leakage, and deformations led to an extensive instrumentation program for monitoring during construction, filling, and long-term operation. The earthfill, foundation, slurry trench, spillway, and river outlet works will all be monitored.

A major concern at Calamus was cracking of the spillway and outlet works conduits from differential settlement, which could cause a potential for piping of silts and fine sand into the conduits. The monitoring of these structures will include use of standard foundation baseplates, structural measurement plates, pneumatic settlement sensors, pneumatic piezometers, and porous tube piezometers. This instrumentation will allow settlement and pore pressure conditions to be closely evaluated in the areas of the structures.

The monitoring of embankment and foundation pore pressures near the slurry trench, within the embankment and foundation, and downstream from the dam will be accomplished with vibrating-wire piezometers, pneumatic piezometers, porous tube piezometers, and observation wells. The ground water conditions were obtained prior to construction and pressure distributions will be obtained and evaluated during filling and steady-state conditions.

The slurry trench cutoff wall will be instrumented immediately upstream and downstream with vibrating-wire piezometers in about four locations along its alinement. In this manner, the effectiveness of the cutoff wall in reducing the foundation pore pressures can be monitored and also compared to the theoretical behavior predicted by the finite element model.

Thermal monitoring has also been used to establish ground water conditions existing at the beginning of construction. Follow-up monitoring will occur following diversion and at intermediate reservoir water surface elevations during the filling process. These data will be compared with information from seepage studies and results of monitoring the piezometers and observation wells.

The drainage system is a major feature for the site, and the monitoring of flows and sediments is a very important consideration. Several manholes along the toe drain and relief well collector and outfall lines were built into the system in order to provide locations for monitoring seepage flows and turbidity. The flows will be measured by interchangeable "v" notch weirs or rectangular weirs depending upon the volume of flow. Vertical flow measurement in relief wells will also be estimated by flow meters. The continuous measurement of turbidity will occur at two points along the toe drain and relief well outfall lines.

Turbidity sensing will include both transmissometer and nephalmeter measurement techniques.

REFERENCES

1. Cedergren, H. R., Seepage, Drainage, and Flow Nets, John Wiley & Sons, Second Edition, 1977.

2. Design of Small Dams, Bureau of Reclamation, Second Edition, 1974.

3. Drainage Manual, Bureau of Reclamation, 1978.

4. Earth Manual, Bureau of Reclamation, Second Edition, 1974.

5. Groundwater Manual, Bureau of Reclamation, 1977.

6. Lambe, William T., and Robert V. Whitman, Soil Mechanics, John Wiley & Sons, Inc., 1969.

7. Mantei, C. L., and D. W. Harris, "Finite Element Seepage Analysis on Reclamation Dams," Preprint 3691, ASCE Annual Convention, Atlanta, Georgia, 1979.

8. Muskeg Engineering Handbook, National Research Council, Toronto, Canada, University of Toronto Press, 1969.

9. Sherard, J. L., et al, Earth and Earth-rock Dams, John Wiley & Sons, 1963.

10. Sherman, W. C., "Filter Concepts," U.S. Army Corps of Engineers.

11. Tracy, F. T., "A Three Dimensional Finite Element Program for Steady-State and Transient Problems," Miscellaneous Paper K-73-3, U.S. Army Engineer Waterways Experiment Station, Vicksburg, Mississippi, 1973.

SEEPAGE IN THE UNSATURATED ZONE: A REVIEW

David B. McWhorter*

The mechanics of flow in partially saturated foundation materials has become an important consideration in the analysis and design of waste impoundments. This paper briefly reviews the procedures whereby the seepage rate, depth of penetration of nonreactive pollutants, and the depth of the wetting front can be calculated when the assumption of one-dimensional flow is justified. A discussion of the effects of macroscale and microscale stratification on lateral spreading is provided. The permeability in each macroscale layer tends toward a constant value dictated by the seepage rate and such layers do not greatly enhance the tendency for lateral spreading. In microscale layers the pressure tends toward a common value, resulting in disparate permeabilities in each layer. This, in turn, causes the permeability tensor to exhibit directional properties that are enhanced relative to those measured at saturation.

INTRODUCTION

Formal scientific investigation of the mechanics of flow of multifluids in porous media probably began with the work of the American phycist E. Buckingham in the first decade of this century (3). Traditionally, research and application of the principles of flow of multifluids in porous media has resided largely in the domain of the soil, petroleum, and hydrologic sciences. In recent years there has occurred an unprecedented explosion of interest in the mechanics of unsaturated porous media, spurred in no small part by problems associated with the containment and disposal of toxic, radioactive, and other hazardous wastes. According to Vick (22), "major steps in understanding the impoundment seepage problem have come about very recently as the result of input from the field of soil science and related areas where movement of water through unsaturated soil has long been of concern." Rather suddenly the movement of solutions and solutes in unsaturated porous media has become a central issue of direct or indirect concern in nearly every sector of society.

This paper presents a review of the salient principles of flow in unsaturated porous media relevant to seepage problems associated with lined waste impoundments. Perhaps the first published work in this specific context was that of McWhorter and Nelson (10,11). Their work, however, consider only one-dimensional seepage into homogeneous, isotropic, and uniformly unsaturated foundation material. While seepage calculations for less idealized situations have been accomplished with computer assisted numerical solutions to the governing equations (4,6,20,13), a generalized description of the

*Colorado State Univ., Engr. Research Ctr., Ft. Collins, CO 80523

effects of stratification, anisotropy, and joints or fractures has not
appeared in the literature. These are commonly encountered features
of foundation materials beneath waste impoundments and further
elucidation of their effects on seepage penetration is warranted.

Further, it became apparent to the author that current literature
dealing specifically with waste impoundment problems contains little
or no discussion of the penetration of the "seepage front" as
distinct from the "wetting front". The seepage front is depth to
which nonreactive pollutants have been carried by the seepage. In
this context, it is the author's opinion that the concept of specific
retention is often erroneously applied, especially in the computation
of the volume of seepage that can be "permanently" stored in the
vadose zone. Misuse of the concept of specific retention, to which
the author may have inadvertently contributed, is discussed herein,
and the circumstance and mechanisms by which permanent storage can
occur are described.

SALIENT FEATURES OF UNSATURATED POROUS MEDIA

The literature dealing with the physics of multifluids in porous
media is extensive and it would be confusing and counterproductive to
attempt a general review here. Rather it is the purpose to bring
forward in this section those features that are particularly relevant
to any discussion of seepage from lined waste impoundments.
Unsaturated porous media, as used herein, will refer to porous media
with pore space occupied simultaneously by air and a liquid phase.
The liquid phase is a solution comprised of water and dissolved
solutes, but will be termed simply as water in the remainder of this
paper.

Pressure Head-Water Content Relation

When water and air coexist in the pores of a porous solid, they
do so at different pressures. With few exceptions, the air pressure
is greater than the water pressure, the water being the wetting phase
and preferentially occupying the smaller pores and recesses of the
medium. As water moves into or out of an element of porous medium,
there is a corresponding change in the volume of air contained in the
element. The viscosity of air, however, is small compared to that of
water, and the resistance to flow of the air is negligible compared to
that for water except for conditons near saturation. The conclusion
of Mukherjee et al. (13) that flow of the air phase is sometimes
important notwithstanding, the pressure of the air in further
developments will be taken constant and equal to zero gage. Under
these circumstances, the water pressure in unsaturated porous media is
negative relative to a zero gage pressure.

Disparate pressures in the air and water phases at points of
contact within the porous matrix are balanced by interfacial forces
arising from surface tension acting on curved interfaces. Equilibrium
of forces acting on the interface between the air and water requires
that the interface become more sharply curved as the water pressure is
increasingly reduced relative to the air pressure. This is
necessarily accomplished by the expulsion of water that, in turn,

permits the interface to be positioned across pore spaces of increasingly smaller dimension. The macroscopic manifestation of these processes is the observed relation between pore-water pressure and water content. Typical relations between pore-water pressure head, h, and volumetric water content, Θ, are shown in Figure 1. The volumetric water content is defined as the volume of water contained in the porous medium per unit volume of porous medium. At complete saturation, the volumetric water content is equal to the porosity, n, of the material.

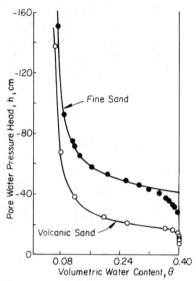

Figure 1. Typical Relationships between Pressure Head and Water Content

The functional relations between h and Θ, denoted hereafter by $h(\Theta)$, vary greatly from material to material. In general, the water content at a particular water pressure will be greater in a fine grained material than in a coarse grained material, other factors being equal. This observation will be seen to have particular relevance in subsequent discussions. The specific function, $h(\Theta)$, that applies for a particular material must be measured. Procedures are described in numerous publications and are well summarized by Corey (5). McWhorter and Nelson (11) provide a means of estimating $h(\Theta)$ in the absence of a measured function.

Darcy's Law in Unsaturated Media. The volume flux of water in unsaturated porous media is given by Darcy's Law which is conveniently expressed as (1)

$$q_j = -K_{ij}\left(\frac{\partial h}{\partial x_j} + \frac{\partial z}{\partial x_j}\right) \quad , \quad i,j = 1,2,3 \tag{1}$$

in which

q_j = component of volume flux in x_j direction

K_{ij} = hydraulic conductivity tensor, a function of pressure head

x_j = jth cartesian space coordinate

z = vertical space coordinate.

h = pressure head (as distinct from total head)

The summation convention for repeated indices applies in eq. 1. Therefore, it represents a short hand notation for a total of three equations, each of which contains three terms on the right side. Equation 1 applies to flow in ansotropic, saturated or unsaturated porous media. The hydraulic conductivity tensor, K_{ij}, is a function of pore-water pressure head in unsaturated media.

For isotropic, unsaturated porous media, the hydraulic conductivity is a scalar function of pressure head and has been widely measured and researched. Figure 2 shows typical relationships $K(h)$ for two sands.

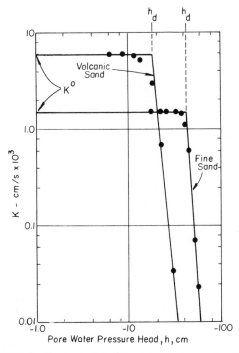

Figure 2. Typical Relationships between K and h

Note that there exists a range of pressure heads over which the hydraulic conductivity is virtually constant. The value of K in this range is equal to the hydraulic conductivity at saturation, K^o. The precipitous decrease of hydraulic conductivity with increasingly negative pressure head is a reflection of the fact that the water content is becoming smaller, causing the dimensions of the space in which water flows to be reduced and the tortuosity of the microscopic flow paths to be increased; the larger pores being desaturated first.

It is especially significant that the $K(h)$ function for the coarser volcanic sand crosses the function for the fine sand. Over the pressure head range of 0 to -20 cm, the hydraulic conductivity of the volcanic sand is greater than that of the fine sand. On the other hand, the hydraulic conductivity of the volcanic sand becomes much less than that of the fine sand at pressure heads more negative than about -20 cm. The fact that the $K(h)$ functions for materials of different texture cross one another is responsible for much of the nonintuitive behavior of seepage in stratified and anisotropic materials. These phenomena are discussed in some detail in a subsequent section.

The hydraulic conductivity function for anisotropic media, $K_{ij}(h)$, has not been thoroughly researched to date. Bear (1) writes

$$(2)$$

$$K_{ij} = K^o_{ij} K_r(h)$$

where K^o_{ij} is the anisotropic hydraulic conductivity at saturation and $K_r(h)$ is the relative hydraulic conductivity. Such a formulation assumes that the relative hydraulic conductivity is a scalar function of pressure head. Further, it predicts that the degree of anisotropy is independent of pressure head; for example, the ratio of horizontal to vertical conductivity is the same at all pressure heads. Recent work by Mualem (12) and Yeh and Gelhar (24) indicates that eq. 2 is not a valid formulation for anisotropy in unsaturated media. These studies show that anisotropy may be a strong function of pressure head owing to microscale stratification that is too small to result in nonhomogenieties on the macro (darcy) scale. Nevertheless, to the author's knowledge, all numerical models for multidimensional flow in unsaturated porous media that attempt to simulate anisotropy do so by assuming the relative hydraulic conductivity to be a scalar function of pressure head and that only the conductivity at saturation is directionally dependent.

STEADY SEEPAGE FROM A LINED IMPOUNDMENT

Steady vertical flow from a lined impoundment overlying a homogeneous foundation material is considered as shown in Figure 3.

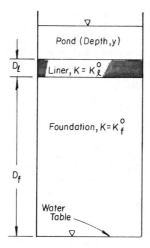

Figure 3. Definition Sketch of Flow System

A problem of this type was first solved by Kisch (9) and later by Zaslavsky (25). A simple analysis (1) shows that the pressure head at the interface between the liner and the foundation material will be negative if

$$\frac{K_f^o}{K_1^o} < (1 + y/D_1) \tag{3}$$

in which K_f^o and K_1^o are the hydraulic conductivities at saturation of the foundation material and the liner material, respectively. The pressure head on top of the liner is y and D_1 is the liner thickness. Situations in which the above inequality is not satisfied are rare because the liner is constructed by compacting a fine textured material, usually clay, so that K_1^o is much less than K_f^o.

It is reasonable to assume that the negative pressure head on the bottom of the liner is insufficient to significantly desaturate the fine textured, highly compacted liner. Therefore, flow occurs through a saturated liner into an unsaturated foundation material in the great majority of cases. The volume flux, q, through the liner is given by

$$q = K_1^o \frac{y + D_1 - h_f}{D_1} \tag{4}$$

in which h_f is the pressure head at the interface between the liner and the foundation. Note that h_f is negative and acts to increase the seepage rate, q.

The pressure head in the unsaturated foundation material will range from zero at the water table to h_f at the bottom of the liner. Integration of the Darcy equation between the water table and the bottom of the liner yields a second independent equation in which the seepage rate q and the pressure head at the interface, h_f, are the only unknowns:

$$D_f = \int_0^{h_f} \frac{dh}{K_f(h) - q} \tag{5}$$

in which D_f is the distance between the water table and the bottom of the liner. Corey (5) and others have shown that eq. 5 requires that

$$q \approx K_f(h_f) \tag{6}$$

in all cases except those in which the water table is very near the surface.

At steady state, the volume flux through the foundation is equal to that through the liner and combining eqs. 4 and 6 yields

$$K_f(h_f) = K_1^o \frac{y + D_1 - h_f}{D_1} \tag{7}$$

The value of h_f for which eq. 7 is satisfied is found by trial and error or graphically. The seepage rate is then determined from either of eqs. 4 or 7.

UNSTEADY SEEPAGE FROM A LINED IMPOUNDMENT

As in the previous section, consideration is restricted to vertical seepage from a lined impoundment overlying a homogeneous and isotropic foundation material. Attention is focused on the estimation of the seepage rate and the depth of penetration of the wetting front.

Seepage Rate. During the initial stage of infiltration into the foundation material, seepage is driven by both gravity and the gradient of pore-water pressure. The pressure head and water content in the wetted zone adjust quickly until gravity is the only driving force and the hydraulic conductivity becomes practically equal to the seepage rate as indicated previously in eq. 6. Following this brief transient period, the seepage rate is constant and equal to that calculated as if the flow were steady. In other words, the pressure head in the wetted zone immediately below the liner adjusts to h_f as calculated from eq. 7. The seepage rate is then calculated from either eq. 4 or eq. 6.

The above described procedure is that used by McWhorter and Nelson (10) to derive their equation for seepage rate. However, they included the effects of headloss through materials placed in the impoundment above the liner rather than simply assuming a pond of depth y as was done herein. For the sake of explicitness and convenience of computation, McWhorter and Nelson used the Brooks-Corey (2) algebraic expressions for the $h(\theta)$ and $K(h)$ relations. Clearly, the Brooks-Corey equations are not an essential feature of the McWhorter-Nelson method.

Penetration of the Wetting Front. Even though the seepage rate, pressure head, and water content immediately below the liner quickly approach constant values, the volume flux, pressure head, and water content at the leading edge of the wetted zone continue to change with time. The leading edge of the wetted zone is characterized by a zone of variable water content, ranging from the antecedent water content in the foundation to θ_f in the wetted zone. The wetting front as used in this paper is defined as the coordinate of a plane located in the zone of variable water content such that an exact volume balance is assured. Specifically,

$$\theta_f z_{wf} - \int_0^{z_{wf}} \theta_i dz = qt \qquad (8)$$

wherein

θ_f = constant, uniform water content in the wetted zone

θ_i = antecendent water content in the foundation

z_{wf} = depth to the wetting front at time, t, measured from liner

For uniform antecedent water content, eq. 8 reduces to the more familiar

$$z_{wf} = \frac{qt}{\theta_f - \theta_i} \qquad (9)$$

Siegel and Stephens (20) compared seepage rate and depth to the wetting front as calculated from the TRUST model with calculations using the above procedures. Their numerical simulations did not artificially restrict flow to the vertical direction as does the McWhorter-Nelson method. Nevertheless, the seepage rates and wetting front depths calculated by both methods were practically identical for isotropic media. The close agreement indicates that lateral spreading of seepage by capillarity in isotropic media was insignificant for the specific situations that were analyzed. Lateral spreading becomes of greater relative importance as the size of the impoundment decreases. Also, lateral spreading was not insignificant in cases where the horizontal hydraulic conductivity was substantially greater than the vertical. Siegel and Stephens concluded that the assumption of one-dimensional flow can introduce substantial error when the foundation is anisotropic, particularly with respect to the wetting front depth.

PENETRATION OF SEEPAGE

It is important to note that, in media containing some antecedent moisture, the location of the wetting front is not the same as the depth to which nonreactive chemicals have penetrated the foundation. Invasion of the foundation material by seepage causes a wetting front to develop and move downward, but the water comprising the wetting front is displaced antecedent water, not seepage water. Smiles, et. al (21) and Hazareth and McWhorter (15) provide data in support of this statement.

When seepage waters invade an unsaturated foundation material, the contact between the seepage water and the antecedent water is not a sharp front. Rather it is a zone in which mixing occurs as the result of hydrodynamic dispersion. Just as for the wetting front, a seepage front is defined as the location of the average depth of penetration of invading seepage waters. The location of the seepage front, z_{sf}, as measured from the liner is given by (21,23,15,19)

$$\int_{0}^{z_{sf}} \Theta \, dz = \int_{0}^{t} q \, dt \tag{10}$$

wherein Θ is the volumetric water content.

It is not difficult to show that, for nonzero Θ_i, z_{sf} is always less than z_{wf} under very general conditions. For the special condition of constant Θ equal to Θ_f in the wetted zone and constant seepage rate, eq. 10 gives

$$z_{sf} = \frac{qt}{\Theta_f} \tag{11}$$

for the position of the seepage (nonreactive chemical) front.

Comparison of eqs. 9 and 11 shows that

$$z_{sf} = (1 - \frac{\Theta_i}{\Theta_f}) z_{wf}, \tag{12}$$

when both Θ_f and Θ_i are uniform. Thus, the depth of penetration of nonreactive may be much less than that of the wetting front at any time. From another view point, the arrival time of seepage waters at a particular depth may lag greatly the arrival time of the wetting front. Chemical constituents in the seepage water that are adsorbed on the solid will lag even the seepage front. Nazareth and McWhorter (15) provide a version of eq. 10 that locates the depth of penetration of linearily adsorbed chemical species.

STORAGE OF SEEPAGE IN THE VADOSE ZONE

Permanent storage of seepage waters in the unsaturated zone has been discussed as a phenomenon that is capable of protecting underlying ground waters from contamination by seepage in arid and semi-arid regions. A common procedure is to regard the difference between specific retention, θ_r, and the antecedent water content, θ_i, as being the volume available for permanent storage per unit volume of foundation material. Such a procedure is not based on physical principles, however.

Specific retention is a rather vague concept for which no unambiguous, commonly held definition exists. The author suspects that its misuse in calculating permanent storage capacity may have resulted because it is often referred to as the water content that cannot be further reduced by gravity drainage. Also, specific retention has been referred to as the residual and irreducible water content, again implying that it represents the capacity of porous media to permanently store water. This interpretation of specific retention is further exacerbated by commonly used algebraic equations for the function $K(\theta)$ that imply that the hydraulic conductivity at specific retention is zero. Putting $K=0$ for $\theta \leq \theta_r$ is only a convenient approximation in many applications and is not an experimentally observed fact. Grismer (8) measured finite values of K in a fine sand for water contents as low as 0.02, far below the water content that would normally be considered as specific retention for that sand.

The variable that dictates whether water is permanently stored in the unsaturated zone is the hydraulic head, not the water content. Water will be stagnant only if the hydraulic head is everywhere the same value. Such will be the case only if the pressure head increases with depth at a rate equal to the change in elevation. This is the condition of hydrostatic equilibrium. It is, of course, necessary that there be no inflow or outflow from the unsaturated zone in order that the water be stagnant. Thus, the important consideration relative to permanent storage of seepage in the unsaturated foundation is the boundary condition that will prevail over the long term. Unless conditions are such that the volume flux of water through the liner eventually becomes zero, there is no possibility of permanent storage of seepage waters in the unsaturated zone, regardless of whether or not the antecedent water content is less than specific rentention.

Even in those cases in which hydrostatic equilibrium will eventually result following cessation of seepage, the volume of seepage waters that can be stored in the unsaturated zone is unrelated to specific retention. The distribution of seepage waters and the volume that can be stored for such cases is profitably examined through eq. 10. Let θ be the difference between the water content after initiation of seepage and the antecedent water content, θ_i. Then

$$\theta = \tilde{\theta} + \theta_i \qquad (13)$$

and eq. 10 is rewritten as

$$\int_o^{z_{sf}} \tilde{\Theta} \, dz + \int_o^{z_{sf}} \Theta_i dz = \int_o^t q \, dt. \tag{14}$$

It is supposed that the antecedent water content distribution, $\Theta_i(z)$, is that which corresponds to the condition of hydrostatic equilibrium. Upon cessation of seepage, the right side becomes a constant, V_s, equal to the cumulative volume of seepage that occurred. Flow in the unsaturated foundation continues, however, as waters redistribute so that hydrostatic equilibrium will again prevail. During this period of redistribution, the seepage front, denoted by Z_{sf}, continues to penetrate deeper into the foundation, the first integral diminishing as Θ becomes smaller and the second integral becoming larger as z_{sf} increases. As hydrostatic equilibrium is approached, Θ vanishes and z_{sf} reaches its maximum value. Equation 14 becomes

$$\int_o^{z_{sf}^*} \Theta_i dz = V_s \tag{15}$$

in which

V_s = cumulative volume of seepage per unit area

z_{sf}^* = maximum depth of penetration of seepage waters.

It is noted from eq. 15 that the seepage waters are permanently stored in the foundation in an interval extending between the liner and a depth z_{sf}. The volume that can be stored in this interval depends upon the antecedent water content, not on specific retention. The largest value that z_{sf} can have is D_f, the depth to the water table. Thus, the maximum volume of seepage that can be stored in the vadose zone is the volume of antecedent water in that zone. Again, it is emphasized that these calculations depend directly on the assumption that the antecedent waters were at hydrostatic equilibrium and that all inflow to the foundation eventually ceases so that equilibrium is again achieved.

EFFECTS OF GEOLOGIC STRATIFICATION

As the wetting front penetrates a stratified foundation material, it must pass through a sequence of layers with different capillary-hydraulic properties. In a particular layer there exists a characteristic pressure head at the wetting front, say h_{wf}. When the wetting front encounters a layer of finer texture, water is readily absorbed. On the other hand, when the wetting front encounters a layer of relatively coarser texture, further penetration of the front is retarded temporarily. These occurrences can be understood by examining the K(h) curves such as those of Figure 2.

Concepts of continuum mechanics requires that the pressure head distribution be continuous across a textural discontinuity. As discussed previously, the hydraulic conductivity of fine textured material is greater than for a coarse textured material at all pressure heads more negative than the pressure head at which the two K(h) functions cross one another (see Figure 2). Thus, the fine layer will usually have a greater hydraulic conductivity that the adjacent coarse layer when the wetting front encounters the textural discontinuity. If the fine layer underlies the coarse layer, water will readily penetrate the fine material because the conductivity is relatively high. When the coarse layer underlies the fine material, water does not readily penetrate because the hydraulic conductivity of the coarse layer is relatively low. In this case, water tends to accumulate immediately above the interface until the pressure head reaches a value for which K(h) of the coarse layer is sufficient to accomodate the flow. The wetting front then penetrates into the coarse layer and moves downward until another layer is encountered.

Once the wetting front has passed through a sequence of strata, the pressure head and water content distributions in that sequence are readily calculated from knowledge of K(h), h(θ), and the seepage rate of q (5). Figure 4 shows a typical distribution of pressure head in a sequence of fine and coarse textured layers.

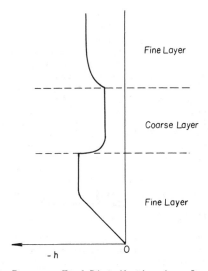

Figure 4. Pressure Head Distribution in a Layered Profile

Note that the pressure heads are constants in each layer, except near the textural discontinuities. This is a reflection of the fact that gravity is the dominant driving force and the pressure heads tend to adjust until the hydraulic conductivities are the same in each layer and equal to the seepage rate.

If the wetting front encounters a layer for which the hydraulic conductivity at saturation is less than the seepage rate, a perched water table will develop. The pore-water pressures in the perched waters will be positive and lateral spreading will occur. The perched zone will spread until total outflow through the perching layer becomes equal to the total inflow to the perched zone. The steady-state areal extent of the perched zone can be roughly estimated from

$$A_p = A_i (q/K^o) \tag{16}$$

where A_p = area of the perched zone

   A_i = area of the impoundment

   q = seepage rate

   K^o = saturated hydraulic conductivity of the perching layer.

Equation 16 results from equating the inflow and outflow from the perched zone under the assumption that the unit area outflow rate is equal to the saturated hydraulic conductivity of the layer on which the water is perched.

EFFECTS OF ANISOTROPY

Anisotropy refers to the directional properties of the hydraulic conductivity at a "point" in the porous medium. A "point" in this context refers to the centroid of a volume element of porous medium that is very small relative to the macroscopic dimensions of the medium, but sufficiently large so that microscopic variations in pore geometry, pressure, fluid velocity, etc. are effectively averaged (1). Anisotropy is uually attributed to stratification on a scale sufficiently small to be effectively averaged in a representatve volume element. If the scale or dimension of the stratification is larger than the dimension of a representative volume element, the medium is considered nonhomogeneous. In other words, macroscopic stratification is treated by analyzing the flow in the individual strata with appropriate conditions imposed at the textural discontinuities. Microscopic stratification is treated by assigning directional properties to the hydraulic conductivity and analysis of flow in individual strata is thus avoided.

Distinguishing between nonhomogeneity and anisotropy on the basis of scale is very important in unsaturated porous media. Flow in unsaturated media with macroscopic stratification results in very different pressures from layer-to-layer as shown in the foregoing section. The different pressures result from a tendency for each layer to adjust so that the hydraulic conductivity is the same. If this same sequence of strata had been analyzed as an anisotropic medium, the pressure would have been considered to be the same in each layer, resulting in greatly different hydraulic conductivities from layer-to-layer.

It is instructive to persue the matter of scale a bit further. A macroscopic, two layered system is considered. The component of flow normal to the bedding will be governed by an average hydraulic conductivity given by

$$K_n = \frac{\delta}{\dfrac{\delta_1}{K_q} + \dfrac{\delta_2}{K_2}} \qquad (17)$$

in which K_n = average hydraulic conductivity normal to the bedding

δ = combined thickness of the layers

δ_1 = thickness of layer 1

δ_2 = thickness of layer 2

K_1, K_2 = hydraulic conductivities of layers 1 and 2 respectively.

The average hydraulic conductivity in the direction parallel to the bedding is

$$K_p = \frac{\delta_1 K_1 + \delta_2 K_2}{\delta} . \qquad (18)$$

The ratio of conductivity parallel to the bedding to that normal to the bedding is

$$\frac{K_p}{K_n} = \frac{(\delta_1 K_1 + \delta_2 K_2)(\delta_1 K_2; \delta_2 K_1)}{\delta^2 K_1 K_2} \qquad (19)$$

If δ_1 and δ_2 are sufficiently small so that the usual devices for measuring the pressure (e.g. a tensiometer) is too large to measure the pressure in the individual layers, then a single value of pressure is assigned to both layers. From Figure 2 it is clear that the hydraulic conductivity of the coarse layer may be greater than, equal to, or less than the conductivity at the fine layer. Data from K(h) curves such as those in Figure 2, used in conjunction with eq. 19 would show that K_p/K_n would first be constant and greater than unity, then decrease to unity, and finally increase rapidly as the pressure head becomes increasingly negative (Figure 5). This qualitative description is entirely consistent with that predicted by Mualem (12) and by Yeh and Gelhar (24). Because the scale of the stratification is too small to permit measurement of pressures in individual strata, this medium would be considered homogeneous and anisotropic.

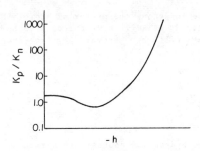

Figure 5. Ratio of K_p/K_n in a Layer Profile

A very different behavior would result if the stratification were on the order of several centimeters or greater. In this case, the media woul be nonhomogeneous, the pressures would be different in each layer, and the conductivities would tend to be much less disparate than in the foregoing. If the flow was steady and normal to a horizontal bedding, the values for K_1 and K_2, averaged over the entire thickness of each layer, would be approximately equal, especially if the layers were very thick. In this case, the K_p/K_n ratio would tend toward unity and be practically independent of pressure.

While the above discussion is far from being completely definitive, it is illustrative of the importance of the scale of the stratification. Treatment of macroscale stratification as if the medium were anisotropic could lead to highly erroneous conclusions as to the depth to the wetting front and the degree of lateral spreading. It is expected that strata with thickness of several centimeters or more must be treated individually if reliable predictions are to result. This question, however, warrants further investigation.

EFFECTS OF FRACTURES

Fractures or joints are important hydrogeologic features, especially in saturated porous media where their permeability often dictates the magnitudes and directions of flow. The effects of such features on flow in unsaturated media comprised of both matric and fracture porosity are more subtle and complicated. Typically, the apertures of fractures and joints are larger than the characteristic dimension of matric pore space and, therefore, desaturate more readily upon a decrease in water pressure. According to Evans and Huang (7) flow in a desaturated fracture occurs in thin films along the fracture face, the thickness of the films being a function of the pore-water pressure.

Given that the water pressures in unsaturated foundation materials are negative and that fractures or joints with apertures greater than the characteristic pore dimension will be desaturated under such conditions, it is expected that fractures and jonts make a

negligible contribution to seepage rates.   Just   as   in   nonfractured
material,   the   seepage   rates   are   controlled   by   the   liner   and   are   not
greatly affected by the propertes of the foundation material.

On the other hand,   fractures   and   joints   could   significantly
affect the distribution of seepage waters in the foundation, depending
upon the orientation of the fractures with respect   to   the   vertical.
Vertical   fractures   will   tend to promote vertical flow by restricting
lateral spreading that would otherwise   occur   by   capillarity.   Flow
from   the porous foundation will not enter an open fracture unless the
pore-water pressure is a least zero gage.   Thus,   vertical   fractures
act   as boundary surfaces, impervious to lateral flow, and effectively
isolate the porous blocks so bounded.

Open, horizontal fractures, or fractures oriented at any   nonzero
angle   relative   to   the   vertical   can be expected to enhance lateral
spreading.   When a wetting front encounters such a fracture, flow will
not   penetrate   into   the   fracture   because   the   water   pressure   is
negative.   In   contrast   to   the   situation   for   vertical   fracture
surfaces,   there   is   a   component   of   the   gravitational driving force
acting   normal   to   the   fracture   surface.   This   driving   force   is
responsible   for   continuing to bring water to the surface, even after
the pressure gradients have been reduced to zero or even   reversed   as
the   water   content   increases   at the surface.   The result is a zone,
extending upward from the fracture surface, in which the water content
and   the   hydraulic   conductivity   have   been   increased.   Thus,   the
tendency for flow in directions tangent to   the   fracture   surface   is
substantially   enhanced.   If the fracture is horizontal, only pressure
gradients act tangent to the interface, and lateral spreading will not
be greatly different than if the fracture did not exist.

SUMMARY AND CONCLUSIONS

Unsaturated   flow   beneath   earthen   lined   waste   impoundments
situated   over   unsaturated   foundation materials occurs because of the
head loss that occurs in the liner.   Penetration   of   the   unsaturated
foundation   by   impoundment   seepage   results   in   two   "fronts";   the
wetting front comprised of   displaced   antecedent   moisture   and   the
seepage   "front"   marking   the average position between the invading
solution and the displaced moisture.   The   seepage   front   moves   more
slowly and always lags the wetting front.

The unsaturated zone beneath a waste impoundment   is   capable   of
permanently   storing   seepage only if the boundary conditions are such
that   all   flow   in   the   unsaturated   zone   will   eventually   cease,
regardless   of the moisture content status of the foundation material.
In those cases where hydrostatic equilibrium will eventually   prevail,
permanent   storage   in   the   unsaturated   zone   is possible and may be
calculated by the methods outlined in this paper.

In   general,   the   effects   of   stratificaion,   anisotropy,   and
fractures   are   such to invalidate simple one-dimensional calculations.
Layered media with individual layers several or more centimeters thick
should   be   treated   as   nonhomogeneous.   Microscopic layering may be
treated by regarding the medium as anisotropic.   Available   numerical

models that calculate flow in unsaturated, anisotropic porous media probably underestimate the effects of anisotropy on lateral spreading. Application of these same models to maeroscopically layered media as if they were anisotropic may over predict the lateral spreading.

APPENDIX - REFERENCES

1. Bear, J., Dynamics of Fluids in Porous Media, American Elsevier, New York, 1972, 764 p.

2. Brooks, R. H., and Corey, A. T., "Properties of Porous Media Affecting Fluid Flow," Journal of the Irrigation and Drainage Division, ASCE, Vol. 92, No. IR2, Proc. Paper 4855, June, 1966, pp. 61-80.

3. Buckingham, E., "Studies in the Movement of Soil Moisture," U.S. Dept. of Agric. Bur. Soils Bull. 38, Washington, D.C., 1907, p. 29-61.

4. Bureau, G., "Design of Impervous Clay Linears by Unsaturated Flow Principles," Proc. 4th Symposium on Uranium Mill Tailings Management, Colorado State University, Fort Collins, 1981, pp. 647-656.

5. Corey, A. T., Mechanics of Heterogeneous Fluids in Porous Media, Water Resources Publications, Littleton, Colorado, 1977, 259 p.

6. Davis, L. A., "Computer Analysis of Seepage and Groundwater Response Beneath Tailing Impoundments," Final Report, National Sci. Foundation, DAR-7917109, April, 1980.

7. Evan, D. D. and Huang, C., "Rate of Desaturation on Transport Through Fractured Rock," in Role of the Unsaturated Zone in Radioactive and Hazardous Waste Disposal; Mercer, J. W., Rao, P.S.C., and Marine, I. W., editors, Ann Arbor Sciences, 1983, pp. 71-79.

8. Grismer, M. E., "Water and Salt Movement in Relatively Dry Soils," Ph.D. Dissertation, Colorado State University, Fort Collins, 1984, 129 p.

9. Kisch, M., "The Theory of Seepage from Clay-Blanketed Reservoirs," Geotechnique, London, England, Vol. IX, 1959, pp. 22-28.

10. McWhorter, D. B. and Nelson, J. D., "Unsaturated Flow Beneath Tailings Impoundments," Journal of Geotechnical Engineering Division, ASCE, Vol. 105, No. GT11, Proc. Paper 14999, November, 1979, pp. 1317-1334.

11. McWhorter, D. B. and Nelson, J. D., "Seepage in the Partially Saturated Zone Beneath Tailings Impoundments," Mining Engineering, Vol. 32, No. 4, April, 1980, pp. 432-439.

12. Mualem, Y., "Anisotropy of Unsaturated Soils," Soil Science Society of America Journal, Vol. 48, 1984, pp. 505-509.

13. Mukherjee, H., Gochnour, J. R., Culham, W. E., and Juvkan-Wold, H. C., "A Numerical Approach to Predict Water Seepage from an Earthen-Lined Evaporation Pond," Proc. 3rd Symp. on Uranium Mill Tailings Management, Colorado State University, Fort Collins, 1980, pp. 233-249.

14. Narasimhan, T. N. and Witherspoon, P. A., "Numerical Model for Saturated-Unsaturated Flow in Deformable Porous Media: I. Theory," Water Resources Research, Vol. 13, No. 3, 1977, pp. 657-664.

15. Nazareth, V. A. and McWhorter, D. B., "Retardation of Adsorbed Chemials in Variably Saturated Flow," Proc. of Symposium on Characterization and Monitoring of the Vadose Zone, Las Vegas, 1983.

16. Neuman, S. P., "Saturated-Unsaturated Seepage by Finite Elements," Journal of Hydraulics Division, ASCE, Vol. 99, No. HY12, 1973, pp. 2233-2250.

17. Neuman, S. P., Feddes, R. A., Brester, E., "Finite Element Analysis of Two-Dimensional Flow in Soils Considering Water Uptake by Roots: I. Theory," Soil Science Society America Proc., Vol. 39, 1975, pp. 231-237.

18. Oster, C. A., "Review of Ground-Water Flow and Transport Models in the Unsaturated Zone," NUREG/CR-2917, PNL-4427, 1982.

19. Raats, P.A.C., "Convective Transport of Ideal Tracers in Unsaturated Soils," Proceedings of the Symposium on Unsaturated Flow and Transport Modeling, Seattle, NUREG/CP-0030, March, 1982, pp. 249-264.

20. Siegel, J. and Stephens, D. B., "Numerical Simulation of Seepage Beneath Lined Ponds," Proc. 3rd Symp. on Uranium Mill Tailings Management, Colorado State University, Fort Collins, 1980, pp. 219-232.

21. Smiles, D. E., Perroux, K. M., Zegelin, S. J., and Raats, P.A.C., "Hydrodynamic Dispersion During Constant Rate Absorpton of Water by Soil," Soil Science Society of America Journal, Vol. 45, No. 4, 1981, pp. 453-458.

22. Vick, S. G., "Planning, Design, and Analysis of Tailings Dams," Wiley-Interscience Publication, John Wiley Sons, New York, 1983, 369 p.

23. Wilson, J.L. and Gelhar, L. W., "Analysis of Longitudinal Dispersion in Unsaturated Flow, 1. The Analytical Method," Water Resources Research, Vol. 17, No. 1, 1981, pp. 122-130.

24. Yeh, T. C. and Gelhar, L. W., "Unsaturated Flow in Heterogeneous Soils," in Role of the Unsaturated Zone in Radioacton and Hazardous Waste Disposal; Mercer, J. W., Rao, P.S.C., and Marine, I. W., editors, Ann Arbor Science, 1983, pp. 71-79.

25. Zaslavsky, D., "Theory of Unsaturated Flow into a Non-uniform Soil Profile," Soil Science, Vol. 97, No. 6, 1964, pp. 400-410.

A CASE HISTORY OF LEAKAGE FROM A SURFACE IMPOUNDMENT

David E. Daniel[1], M. ASCE, Stephen J. Trautwein[1],

and David C. McMurtry[2]

ABSTRACT

A manufacturing company constructed a 48-acre (19 hectare), unlined surface impoundment at a site in northern Texas to store wastewater. The site is underlain by unsaturated, alluvial soils. The water table is at a depth of 390 ft (119 m). The upper 50 ft (15 m) of subsoil is a clayey material.

An investigation of groundwater conditions beneath the pond, beginning in 1980, showed that a large mound of contaminated groundwater was perched above the aquifer. A number of finite element analyses were performed in which the hydraulic conductivities of the soil strata, as well as other soil properties, were adjusted until the calculated dimensions of the mound of contaminated water matched the known dimensions. The hydraulic conductivities that were needed to achieve good agreement between computed and measured groundwater conditions were, for some of the strata, several orders of magnitude higher than values obtained from laboratory permeability tests. A limited number of field permeability tests at shallow depth yielded hydraulic conductivities which agreed fairly well with the values from the finite element analyses. This case history provides a good example of an instance where use of laboratory permeability tests would lead to grossly misleading projections of the magnitude of leakage from a surface impoundment.

INTRODUCTION

In 1980, the state of Texas directed a large chemical manufacturing company to investigate groundwater conditions beneath various wastewater storage ponds at one of its plants in northern Texas. Of main concern was an unlined, 48 acre (19 hectare) pond that had been in operation since 1960. The wastewater stored in the impoundment contained total organic carbon (TOC) concentrations of several thousand mg/L and high levels of dissolved inorganic compounds. The

---

[1] Assistant Professor and Graduate Student, respectively, University of Texas, Department of Civil Engineering, Austin, Texas.

[2] Civil Engineer, Underground Resource Management, Inc., Austin, Texas.

organic chemicals in the wastewater included acetate and benzene. Sodium, sulfate, and manganese were the most significant inorganic constitutents. The depth of water in the pond was less than 3 ft (90 cm), and the main purpose of the impoundment was to promote evaporation of the wastewater by spreading it over a relatively large area. The natural groundwater table at this site is 390 ft (119 m) below the surface. The site is located in an arid region where the average yearly precipitation is 20 inches and the estimated potential evaporation is 73 inches per year.

In response to a directive to investigate groundwater conditions, two deep monitor wells were installed and were screened over a depth interval of 300 to 430 ft (91 to 131 m). The regional aquifer is unconfined and located at a depth of approximately 390 ft (119 m) with a saturated thickness of approximately 20 ft (6 m) at the plant site. The aquifer is the only permanent source of fresh water in the area. Groundwater samples were recovered from the two monitor wells and were analyzed for TOC and manganese content, among others. One well showed no contamination, but the well located closest to the surface impoundment showed an elevated manganese concentration. It was suspected that contaminated water from near the ground surface had somehow made its way down the well bore (perhaps as the result of an improperly sealed hole above the screened section), but there was no way to prove this without installing additional monitor wells.

Based upon the contamination found in the initial survey, a more detailed investigation was launched. However, nearly everyone involved in the project was convinced that there was not likely to be any adverse impact on groundwater quality based on the fact that the evaporation pond was underlain by 50 ft (15 m) of clayey soils. Accordingly, the initial stage of investigation was aimed at simply confirming that contaminants did not penetrate to significant depths.

In this paper, the investigative work for the project is described chronologically.

THE IMPOUNDMENT AND SUBSOILS

A plan view of the site is shown in Fig. 1 along with locations of various monitor wells and other instruments to be discussed later. The pond was built above grade with dikes 5 to 10 feet high.

The site is underlain by alluvial soils that were broadly categorized by such terms as "sand", "clay", "sandy clay", and similar terms. A typical profile is shown in Fig. 2. The soils above the water table have degrees of saturation of 20 to 70 percent outside the area influenced by the impoundment. Because the ground surface is flat and the soil strata do not dip significantly and are continuous over relatively large areas, it was assumed that moisture conditions outside the area of the pond at the time of the investigation were the same as the moisture conditions beneath the impoundment at the time that the impoundment was constructed.

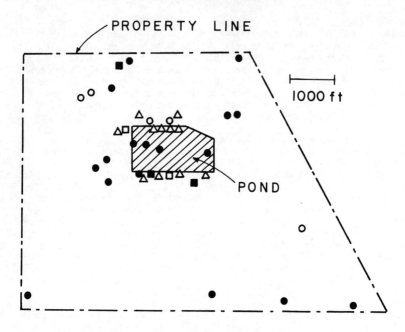

Figure 1. Plan View of Site.

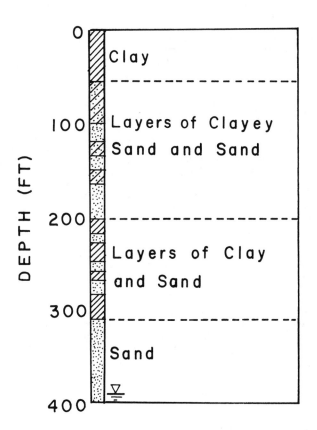

Figure 2. Generalized Soil Profile

The soil located immediately beneath the impoundment is a clayey soil with a thickness of approximately 50 ft (15 m). The average liquid and plastic limits of the clayey soil are 41 and 18 percent. respectively. Sieve analyses indicated that 80 to 98 percent of the soil by dry weight passed the No. 200 sieve. Laboratory permeability tests of an unknown type were performed prior to construction of the pond and yielded coefficients of permeability (k), or hydraulic conductivities, between 1×10^{-8} and 1×10^{-9} cm/sec. Few additional laboratory test data are available.

INITIAL STUDIES

Hand Calculations

The initial plan was to investigate the shallow soils beneath the evaporation ponds. This plan was based on two sets of calculations. In the first set, the hydraulic conductivity of the layer of clayey soil was assumed to be 1×10^{-8} cm/sec, and a unit hydraulic gradient was assumed. The distance that the pond water migrated over a period of 20 years was computed assuming one-dimensional seepage and using a range of values for effective porosity. The calculations indicated that the water should not have migrated more than a few feet.

Numerical Analyses

Because the soil into which the pond water was migrating was initially unsaturated, it seemed appropriate to perform a more rigorous analysis of water flow into an initially unsaturated porous medium. Although closed-form analysis was possible, it was thought that numerical analysis might eventually be needed to analyze moisture migration over the full depth of unsaturated soils under a range of assumed boundary conditions. For these reasons, a numerical model was used to analyze the relatively simple case of moisture movement into an initially unsaturated stratum of clayey soil. The model that was used is a finite element code called DUNSAT (1). Assumptions were made regarding the moisture-suction curves for the various substrata, and relationships between hydraulic conductivity and degree of saturation were assumed. A range in values of hydraulic conductivity at full saturation was assumed, with values for the uppermost layer falling in the range of 1×10^{-7} to 1×10^{-8} cm/sec. The initial conditions were established theoretically by imposing a known moisture content (and hence controlled suction) near the ground surface and determining the long-term profile of moisture content vs. depth. No attempt was made to account for the cyclic nature of near-surface moisture conditions because variations in precipitation/evapotranspiration were not thought to influence moisture conditions to significant depths. The moisture-suction curves were adjusted until the model predicted the moisture content profile actually measured in samples collected outside the area influenced by pond seepage. Once the initial (before ponding) suctions and moisture contents were established, the computer program was used to calculate the changes in moisture content that would be expected over a 20 yr. period of ponding. The analyses indicated that the pond water should not have penetrated to depths of more than 10 to 40 ft (3 to 9 m), depending on the hydraulic conductivities at full saturation that were assumed.

As expected, more rigorous analysis showed about the same results as
simple hand calculations had previously, i.e., given the low hydraulic
conductivities that were assumed for the stratum of clayey soil that
underlies the pond, one would not expect the pond water to have
migrated very far into the subsoil.

Field and Laboratory Investigation

Twenty exploratory borings were drilled to investigate the actual
groundwater conditions. All but two of the borings were drilled to a
depth of 40 ft (12 m). Two borings were drilled to a depth of 100 ft
(30 m). One of the deep borings was drilled near the center of the
pond and the second was drilled outside of the area affected by the
pond to aid in establishing baseline (before pond) moisture conditions.
Pressure-vacuum lysimeters were installed in the boreholes to withdraw
groundwater samples from saturated or unsaturated soils.

A comparison of moisture content profiles from borings drilled
inside and outside the pond area indicated that degree of saturation
was not appreciably higher for the soil located below a depth of about
40 ft (12 m). The moisture content profile that was measured in the
field seemed to confirm the results of the first set of analyses, viz.,
contaminated groundwater was confined to the upper stratum of clayey
soil. However, chemical analyses of soil and water samples from
beneath the pond indicated the presence of contaminants to depths of
at least 100 ft (30 m).

At first it was difficult to accept the indications that pond
water had seeped through the upper layer of clay. Drilling and
sampling procedures were reviewed to check the possibility that they
may have been the source of contamination. Once it was established
that there were no serious flaws in the sampling procedures, and that
the contaminants detected in the groundwater did indeed appear to be
from the surface impoundment, the scope of the investigation expanded.

SECOND PHASE OF INVESTIGATION

If the contaminants did migrate to depths of 100 ft (30 m) or more
the average hydraulic conductivity of the clayey soils that underlie
the impoundment must have been greater than the values that had been
assumed previously. Accordingly, the initial thrust of the second
stage of investigation was to evaluate the hydraulic conductivity of
the clayey soil in more detail.

Permeability Tests

Twenty eight laboratory permeability tests were performed on 3 in.
(7.5 cm) diameter, undisturbed samples, using fixed-wall permeameters.
Tap water was used for most tests. Pond water was used for 2 tests,
and the pond water had no detrimental effect upon hydraulic conductivity. Backpressure was not used in the tests. The measured
hydraulic conductivities were between 2×10^{-5} and 1×10^{-9} cm/sec, with
an average of 2×10^{-6} cm/sec.

Four field permeability tests were conducted in shallow boreholes drilled into the stratum of clayey soil. The tests were conducted in 9 in. (23 cm) diameter augered holes that extended to depths of less than 20 ft (6 m). A screened section 5 ft (152 cm) long was backfilled with sand, and a 2 in. (5 cm) diameter pipe (annular space backfilled with bentonite) extended to the ground surface. Falling head tests yielded hydraulic conductivities of 3×10^{-6}, 3×10^{-6}, 3×10^{-4}, and 2×10^{-4} cm/sec. The method used to calculate k is described in Ref. 2. The average field value (1×10^{-4} cm/sec) is almost 2 orders of magnitude higher than the average laboratory value. A direct comparison of laboratory and field values of hydraulic conductivity for soil from the same boring and depth range (Table 1) shows that field values were 100 to 10,000 times greater than laboratory values. One reason for the difference is that the field test results are more indicative of the horizontal rather than the vertical hydraulic conductivity. However, the most likely explanation for the large difference is that the laboratory tests did not account for seepage paths such as cracks and root holes. Root holes were observed in soil samples to depths of up to 40 ft (12 m).

Analysis

After updated values of hydraulic conductivity had been obtained, the one-dimensional finite element anlayses were redone to estimate the extent of migration from the pond. The values of hydraulic conductivity for full saturation (k_{sat}) that were assumed are tabulated in Table 2. The assumed moisture-suction curves are shown in Fig. 3, and the assumed hydraulic conductivity functions are shown in Fig. 4. The curves shown in Figs. 3 and 4 were constructed using published data for soils with similar gradational characteristics and were adjusted until the computed and measured moisture content profiles matched reasonably well for the before-ponding case.

The results of the analysis of moisture migration after 20 yrs of ponding are shown in Fig. 5. The analysis showed that after 20 yrs of ponding water at the surface, the maximum depth of moisture content increase was approximately 330 ft (100 m). Interestingly, the program predicted that the subsoil would be saturated to a depth of 50 ft (15 m), but that an unsaturated zone would exist between depths of 50 and 150 ft (15 to 46 m). It appeared that the surficial layers of clayey soil impeded the downward movement of water, and that the underlying stratum of clayey sand could conduct all the water that was supplied to it from above while maintaining a partially-saturated condition. The calculations showed, however, that a deep zone of perched water began forming when seepage encountered the sandy clay at a depth of 220 ft (67 m) and began mounding on top of this layer of relatively low hydraulic conductivity material.

The analyses were carried out to larger times to project what might happen if the pond continued to operate or if the impoundment were closed. In either case, the calculations indicated that contaminated water would reach the underlying aquifer in another 20 to 30 yrs. Cutting off the source of water at the surface did little to reduce the time until impact of the mound of contaminated water on the aquifer.

TABLE 1. SUMMARY OF RESULTS FROM FIELD AND
          LABORATORY PERMEABILITY TESTS

Borehole Number	Hydraulic Conductivity (cm/sec)	
	Field Test	Lab. Tests*
1	3×10^{-6}	8×10^{-8} 2×10^{-8}
2	3×10^{-5}	1×10^{-8}
3	3×10^{-4}	2×10^{-8}
4	2×10^{-4}	---

*Performed on undisturbed samples obtained from the elevation of
the center of the screen that was later installed for field
permeability testing.

TABLE 2. VALUES OF HYDRAULIC CONDUCTIVITY OF SATURATED SOIL
          (k_{sat}) USED FOR ONE-DIMENSIONAL ANALYSES IN SECOND
          STAGE OF THE INVESTIGATION

Layer	Depth Interval		Soil Type	k_{sat} (cm/sec)
	Feet	Meters		
1	0 - 20	0 - 6	Silty Clay	1×10^{-4}
2	20 - 50	6 - 15	Silty Clay	1×10^{-6}
3	50 - 70	15 - 21	Sandy Clay	1×10^{-5}
4	70 - 140	21 - 43	Clayey Sand	1×10^{-4}
5	140 - 170	43 - 52	Sand	1×10^{-3}
6	170 - 220	52 - 67	Clayey Gravel	1×10^{-4}
7	220 - 300	67 - 91	Sandy Clay	1×10^{-7}
8	300 - 390	91 - 119	Sand	1×10^{-3}

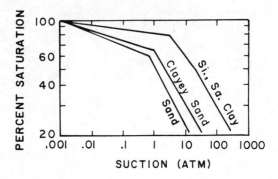

Figure 3. Moisture-Suction Curves that Were Assumed.

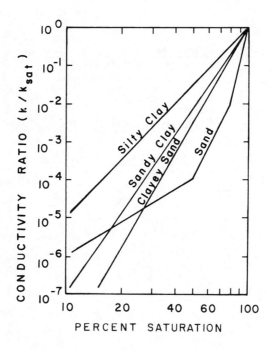

Figure 4. Assumed Curves of Conductivity Ratio Vs.
Percent Saturation.

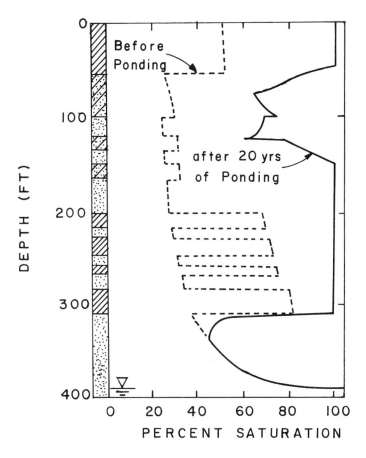

Figure 5. Profiles of Percent Saturation for the Before Ponding
Case and the Calculated Conditions after 20 yrs of
Ponding.

The engineers working on this project realized that a number of simplifying assumptions had been made in the analyses, and that predicting transport times with any reasonable degree of accuracy is a very difficult task. Those working on the project simply concluded from the analyses that the aquifer may be impacted within the next several decades, if not sooner.

THIRD PHASE OF INVESTIGATION

Field Investigation
The next stage of investigation involved additional field work to determine the actual extent of groundwater contamination. Monitor wells and lysimeters were installed at various depths throughout the profile and over the entire plant site (Fig. 1). More detailed moisture content profiles beneath the ponds were constructed and additional chemical analyses were performed on groundwater samples.

The field studies confirmed the presence of a perched mound of contaminated groundwater between depths of 140 and 300 ft (43 and 91 m). The size of the mound was much larger than anticipated; the field data indicated that the mound had a radius of nearly a mile (1.6 km), as shown in Fig. 6.

One-Dimensional Analyses
Although it was becoming clear that seepage from beneath the pond was not one-dimensional, it was decided to complete the one-dimensional analyses by refining the values of hydraulic conductivity for saturated soil until a match was obtained with the measured moisture content profile beneath the center of the pond. The resulting hydraulic conductivities were similar to those which had been used previously, but were slightly larger.

Two-Dimensional Analyses
The computer program was also used for three-dimensional analyses. Axi-symmetric conditions were assumed based upon the nearly symmetric shape of the groundwater mound that formed beneath the pond (Fig. 6).

The first set of two-dimensional analyses were performed using the soil properties that were obtained from the last set of one-dimensional analyses. The two-dimensional analysis predicted the correct moisture content profile beneath the center of the pond but indicated very little lateral spreading. This was contrary to the field observations.

Several additional two dimensional analyses were performed, with adjustments made to the values of hydraulic conductivity at saturation in the horizontal and vertical directions with each new trial, until the predicted shape of the groundwater mound matched the actual shape reasonably well. After several trials, a reasonable match was obtained between the predicted shape of the groundwater mound and the actual shape. In order to obtain reasonable agreement, the hydraulic conductivities of the upper layers had to be increased and the ratio

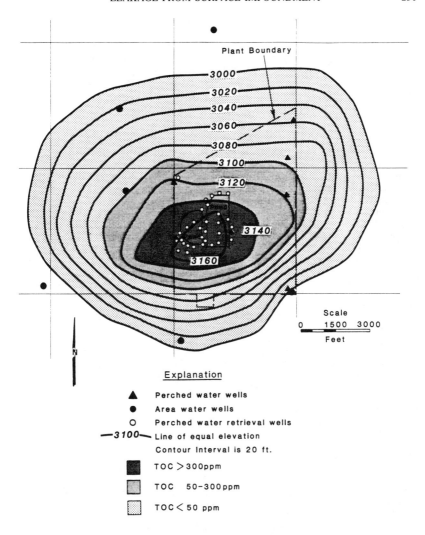

Figure 6.   Locations of Retrieval Wells with Contours of Elevation
            of Top of Mound of Perched Groundwater and Contours of
            Equal TOC Concentrations.

of the horizontal to vertical hydraulic conductivity had to be increased from 1 to a value of 10 in the layers between 140 and 200 ft (43 and 61 m). Several other approaches were tried, such as uniformly increasing all hydraulic conductivities, but they did not result in a computed mound that matched the shape of the actual mound. Increasing k of the layer at 300 ft (91 m) was tried, but the resulting mound was not nearly as large as the actual mound. A summary of the final hydraulic conductivities for saturated soils that were used in the two-dimensional analyses is given in Fig. 7.

Miscellaneous

Contaminant Transport. There was doubt that the contaminants in the groundwater would move at the same velocity as the transporting groundwater. Batch and column tests were performed in the laboratory to study attenuation of contaminants in the soils found at the site. It was found that significant attenuation of the metals would be expected (due primarily to precipitation in a reducing environment) but that the organics were not attenuated significantly. Thus, the conservative assumption made at the outset that some of the contaminants would probably move at the same velocity as the transporting water was ultimately found to be valid.

Contamination in Deep Monitor Well. The first sign of problems at this site was the sampling of a monitor well that tapped the regional aquifer beneath the site at a location close to the surface impoundment. The sampled water had elevated manganese content, but it was originally thought that contaminants had migrated from the surface through a poorly sealed well. Because the top of the screened section actually intersected the bottom of the mound of perched contaminated groundwater at a depth of 300 ft (91 m), the screened section actually intersected the bottom of the mound of contaminated groundwater. The contaminants apparently migrated down the screened part of the well, not along the sealed portion above the screened zone. Monitor wells have consistently shown the aquifer to be uncontaminated.

Numerical Analyses. No one took the numbers from the computer output too seriously, because there were numerous simplifying assumptions that had been made, but once reasonable values of hydraulic conductivity were used, the computer program consistently pointed to the correct patterns of moisture movement. The main value of the computer analyses was the added insight that if provided to what had happened to the groundwater at this site over the previous 20 yrs.

Mass Balance Calculations. Attempts were made early in the project to perform a mass balance on the pond to determine the amount of leakage, but plant records indicating the amount of wastewater that had been pumped into the pond could not be located. However, near the end of the project, when the severity of the problem was appreciated, plant records were located to indicate the historical rates of pumpage of wastewater into the pond. The evaporative losses were estimated, and the difference between the amount of wastewater pumped into the pond and the amount lost to evaporation was assumed to be the amount of leakage. This turned out to be 17 ft (5.2 m) of seepage

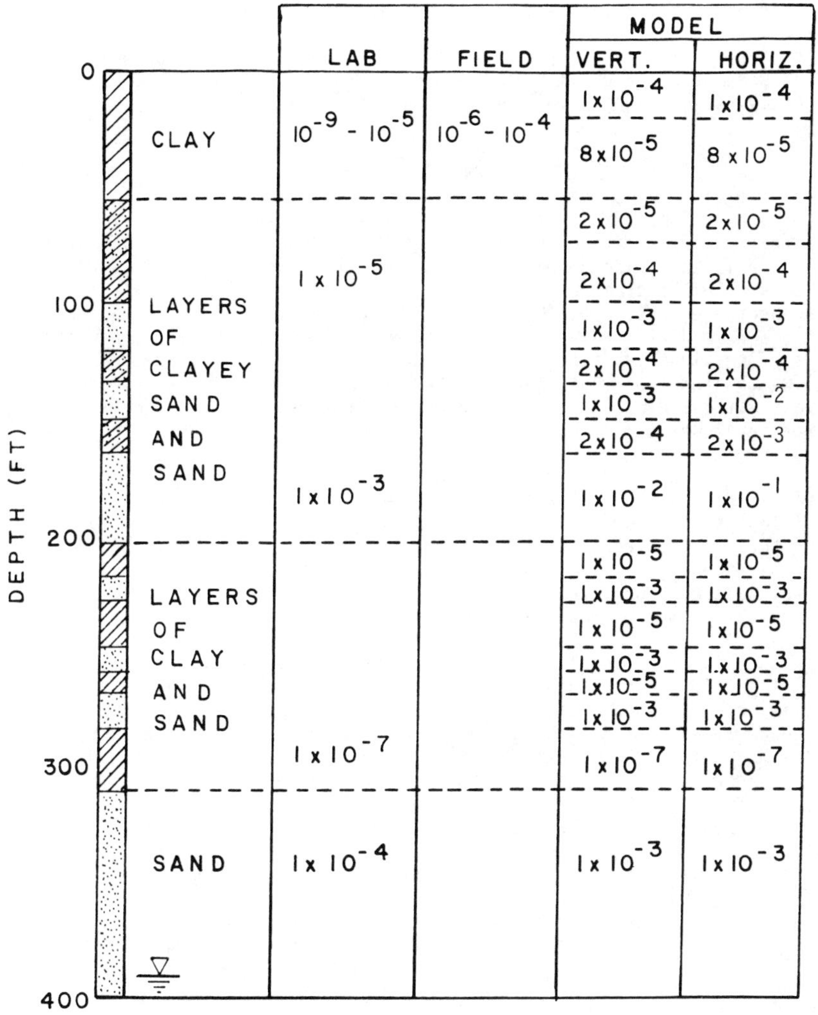

DEPTH (FT)

		LAB	FIELD	MODEL VERT.	MODEL HORIZ.
0	CLAY	$10^{-9} - 10^{-5}$	$10^{-6} - 10^{-4}$	1×10^{-4} / 8×10^{-5}	1×10^{-4} / 8×10^{-5}
100	LAYERS OF CLAYEY SAND AND SAND	1×10^{-5} / 1×10^{-3}		2×10^{-5} ; 2×10^{-4} ; 1×10^{-3} ; 2×10^{-4} ; 1×10^{-3} ; 2×10^{-4} ; 1×10^{-2}	2×10^{-5} ; 2×10^{-4} ; 1×10^{-3} ; 2×10^{-4} ; 1×10^{-2} ; 2×10^{-3} ; 1×10^{-1}
200 – 300	LAYERS OF CLAY AND SAND	1×10^{-7}		1×10^{-5} ; 1×10^{-3} ; 1×10^{-5} ; 1×10^{-3} ; 1×10^{-5} ; 1×10^{-3} ; 1×10^{-7}	1×10^{-5} ; 1×10^{-3} ; 1×10^{-5} ; 1×10^{-3} ; 1×10^{-5} ; 1×10^{-3} ; 1×10^{-7}
400	SAND	1×10^{-4}		1×10^{-3}	1×10^{-3}

Figure 7. Values of Hydraulic Conductivity (cm/sec) Used in the Final Two-Dimensional Analyses.

loss per year. A calculation of the size of the mound for this rate of leakage yielded dimensions that were very close to those indicated by the finite element analyses. If the data needed for a water balance had been obtained earlier in the project the extent of groundwater contamination might have been appreciated earlier.

REMEDIAL ACTION

Thirty-one retrieval wells have recently been installed at this site (Fig. 6) to recover most of the contaminated groundwater. The wells were placed near the center of the mound where the hydraulic heads were the greatest and the TOC levels were the highest. The wells are screened over a depth range of 140 to 275 ft (43 to 84 m) and are to be pumped at a rate of 8 gpm (143 m^3/day) per well. The impoundment is being closed as alternative surface discharge permits are granted. A wastewater treatment plant has been constructed to treat process wastewater and wastewater retrieved from the wells. Additional retrieval wells may be needed later.

The calculated hydraulic conductivities from the known pumpage rates that have been applied to the retrieval wells are in the range of 10^{-3} to 10^{-4} cm/sec; these represent gross averages over the depth of the screened portion of the well (140 to 275 ft, or 43 to 84 m). This range is consistent with the values of hydraulic conductivity that were used when the finite element program was used to predict the known size of the groundwater mound.

It is expected that the retrieval wells will be in operation for at least 10 years. The cost of the cleanup is on the order of $10 million. It has been estimated that the retrieval well system will reduce the amount of TOC reaching the aquifer by 45%. Actual impact on the aquifer is not expected for an additional 20 years. Because the water in the aquifer is flowing down gradient very slowly, the contaminant plume is not expected to reach any major wells for at least several decades. Additional studies are currently being conducted to better access the effects of aquifer impact. Also, monitor wells have been installed into the aquifer, down gradient from the ponds, in order to detect contamination. Meanwhile, the fate of the organics that will eventually reach the aquifer is under study; these efforts include a study of dilution and degradation.

CONCLUSIONS

A large surface impoundment was constructed at a site in northern Texas in 1960. The natural groundwater table at the site is 390 ft (119 m) below the surface at the top of the primary aquifer for the area. The impoundment contained wastewater with a number of organic and inorganic constituents. The organics were largely unattenuated by the subsoils, based on laboratory adsorption tests. Laboratory measurements of hydraulic conductivity performed prior to construction of the pond showed values of 10^{-8} to 10^{-9} cm/sec in a layer of clayey soil that formed the upper 50 ft (15 m) of subsoil. Little leakage was expected, based on the low hydraulic conductivities.

An investigation of the groundwater conditions at the site, beginning in 1980, showed that a large mound of contaminated groundwater was perched above the aquifer. The mound ranged from a depth of 140 ft to over 300 ft (43 to 91 m) and extended laterally as much as 1 mile (1.6 km) from the 48-acre (19 hectare) pond. It was interesting that the soil from the ground surface to a depth of 50 ft (15 m) was saturated from pond water, but the soil from 50 to 140 ft (15 to 43 m) was not saturated. Beneath the unsaturated zone, the soils were saturated to a depth of over 300 ft (91 m). The unsaturated zone, sandwiched between two saturated zones that contained contaminated water, led to much confusion in the initial stages of the study and even led to some incorrect conclusions about the extent of contamination.

Finite element analyses correctly predicted the presence of an unsaturated zone sandwiched between two saturated zones once reasonable values of hydraulic conductivity were used. A number of analyses were performed, and hydraulic conductivities (as well as other soil parameters) were varied until the predicted pattern of moisture migration matched the observed pattern. The hydraulic conductivities that were necessary to match the predicted and measured patterns of moisture movement were much higher than laboratory permeability tests had indicated (Fig. 7). If one were to predict the extent of groundwater contamination at this site on the basis of available laboratory permeability test results, the predictions would be grossly in error. In order to improve our ability to predict leakage rates from surface impoundments, the single greatest need is to improve our ability to determine the hydraulic conductivity of earth materials over relatively large areas.

ACKNOWLEDGMENTS

The batch-adsorption tests were performed under the direction of Prof. Howard Liljestrand at The University of Texas. Prof. Randall Charbeneau provided input on contaminant transport and design of the retrieval well system. Michael Cooper was the technical director of the project for Underground Resource Management. Charles Maulden directed much of the field work.

REFERENCES

1. Daniel, D.E., "Moisture Movement in Soils in the Vicinity of Waste Disposal Sites," Dissertation Presented to the Faculty of the Graduate School of The University of Texas at Austin in Partial Fulfillment of the Requirements for the Degree of Doctor of Philosophy, 1980, 235 p.

2. Hvorslev, M.J., "Time Lag and Soil Permeability in Ground Water Ovservations," Bulletin No. 36, U.S. Army Waterways Experiment Station, Vicksburg, Mississippi.

ASSESSING SEEPAGE AT UNCONTROLLED HAZARDOUS WASTE SITES

By Brad J. Berggren,[1] A.M. ASCE,
Lawrence A. Holm,[2] A.M. ASCE, and
James R. Schneider,[3] M. ASCE

ABSTRACT: Many uncontrolled hazardous waste sites are underlain by soil contaminated with a wide variety of chemical compounds. Infiltration often carries these compounds into groundwater aquifers where they form plumes of contaminated water. Each compound leaches from the soil and migrates at different rates, forming what may be considered to be a variety of plumes. When comparing potential cleanup alternatives that include groundwater extraction, it is necessary to estimate the concentrations of contaminants that will be withdrawn from extraction wells and the variation in concentrations of the contaminants with time. It is also important to be able to compare the relative costs of soil remedial actions (e.g., capping and/or excavation) with costs for groundwater remedial actions such as extraction and treatment. This paper presents a conceptual framework for estimating the concentration of organic contaminants in a groundwater plume over time. The basis of this framework is the application of adsorption-desorption principles together with chemical mass-balance considerations for organic contaminant depletion. The approach considers the concentration of organic contaminants in the partially saturated soil above the groundwater level, as well as contaminants already in the groundwater at startup of extraction. The approach is meant for use in a pragmatic manner, as an aid to considered judgment; it is not proposed as a strict analytical model. It is hoped that presentation of the concepts will stimulate discussion and applied research on this important topic.

INTRODUCTION

At many hazardous waste sites, a wide variety of chemicals have become mixed with near-surface soils by seepage, by spillage, or by intentional dumping. Precipitation and infiltration often convey these chemicals into the groundwater, resulting in formation of a plume of contaminated groundwater moving through an aquifer. When comparing alternatives for groundwater cleanup, several questions relating to the projected removal rate of contaminants need

[1]Geotechnical Engineer, CH2M HILL, Milwaukee, Wisconsin.
[2]Geotechnical Engineer, CH2M HILL, Redding, California.
[3]Geotechnical Engineer, CH2M HILL, Portland, Oregon.

to be addressed in order to rationally compare cleanup alternatives:

1. How long must groundwater extraction be continued to achieve a specified degree of cleanup in the aquifer?

2. During what portion of the extraction period must the withdrawn groundwater be treated to meet discharge criteria?

3. What concentrations of various contaminants will the treatment system have to handle, and how will treatment needs vary over time?

4. What effect does capping or partial excavation of contaminated soil have on the preceding three issues?

To address these issues, it is necessary to estimate the variation in concentration of contaminants in the groundwater with time. This paper presents concepts that may be used to develop reasonable answers to these questions for organic compounds. The concepts presented are intended to be practical tools and are not intended to serve as a mathematical model of contaminant migration in the ground.

NATURE OF THE PROBLEM

To illustrate the difficulties associated with such estimates, consider a hypothetical uncontrolled hazardous waste site with features typical of uncontrolled hazardous waste sites: an open, unlined lagoon; a gravelled yard where numerous drums containing a variety of waste chemical compounds were stored; and/or a number of abandoned tanks where compounds were blended or stored. Soil at the site may have been contaminated by seepage, by spills, or perhaps by intentional, random dumping.

After these compounds are in the soil, what happens to them? Consider the section shown in Fig. 1. In this example, the unsaturated zone contains soil contaminated with a variety of chemicals. Water infiltrating the contaminated soil in this "source zone" picks up some of the compounds in solution and carries them downward to the saturated zone. On reaching the saturated zone, the contaminants in solution migrate in response to the local groundwater gradient, slowly mixing, dispersing, and diffusing with uncontaminated groundwater moving beneath the site. This plume moves through the aquifer, contaminating the aquifer as some of the compounds are adsorbed and desorbed to and from the soil skeleton.

The rate at which a contaminant plume moves depends on the aquifer characteristics and on the chemical characteristics

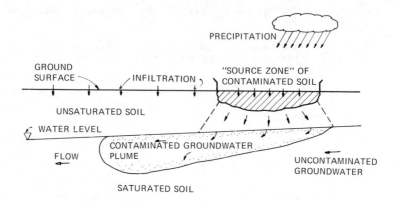

FIG. 1 — Section Through Hypothetical Contaminated Groundwater Plume

of each compound in the plume. Compound movement rates are
affected by each compound's affinity for solids within the
soil skeleton. A result of these differing affinities is
that when a site is contaminated by more than one chemical,
multiple plumes may exist simultaneously, each moving at a
different rate. Fig. 2 presents four plume configurations
for comparison. In Fig. 2(a), contours of equal concentration of a low mobility compound such as DDT are shown
after an arbitrary amount of time since initial site contamination. Note that the center of the plume is still
located close to the site. Fig. 2(b) shows contours of
equal concentration of a moderate mobility compound such as
trichloroethene (TCE) after the same period of time.

In this case, the center of the plume has begun to move
away from the site. Fig. 2(c) shows contours of equal concentration for a high mobility compound such as methylene
chloride after the same period of time. In this case, the
zone of highest concentration has already moved well away
from the site. Fig. 2(d) superimposes the lowest concentration contour of each compound to illustrate this concept
of multiple plumes.

If consideration is being given to removing the contaminated groundwater by construction of an extraction well at
point A, what chemicals will be found in the well discharge, and in what concentrations? As can be seen in
Fig. 2(d), at first methylene chloride will appear in the
well discharge. Since the well will also withdraw uncontaminated water from the aquifer, the concentration of
methylene chloride in the well discharge will be less than

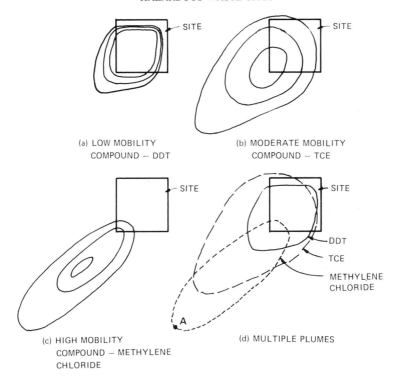

(a) LOW MOBILITY
COMPOUND – DDT

(b) MODERATE MOBILITY
COMPOUND – TCE

(c) HIGH MOBILITY
COMPOUND – METHYLENE
CHLORIDE

(d) MULTIPLE PLUMES

FIG. 2 – Groundwater Plumes; Hypothetical Contours of Concentration

that in the plume adjacent to the well. As the plume continues to move toward the well, the concentration of methylene chloride will increase, and eventually TCE will begin
to show up in the well discharge. As the concentration of
TCE increases, the concentration of methylene chloride may
continue to increase or it may peak and begin to decrease.
Still later, DDT will begin to show up in the well discharge, perhaps while the concentration of methylene chloride is decreasing and near the peak concentration of TCE.
This is depicted schematically in Fig. 3. Clearly, estimation of the well discharge contaminant concentration is
difficult at a site where perhaps 50 assorted organic pollutants are present.

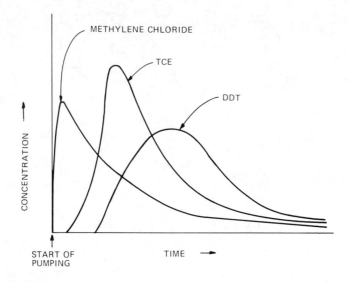

FIG. 3 — Concentration Variation with Time for Several Contaminants

PRACTICAL APPROACH

In order to stimulate thought and discussion, this paper
presents an approach recently applied to a complex site to
address the questions outlined previously. This approach
is based on the application of sorption principles and mass
balancing.

SORPTION PRINCIPLES

Information regarding adsorption-desorption phenomena has
been developed through work in several disciplines. Currently, this is a dynamic area of science with considerable
effort directed toward extending the present somewhat limited understanding of the processes involved.

Several general equations and approaches have been developed to describe the sorption process; however, application
of these equations with any degree of accuracy is often
limited. The approach outlined in this paper uses a simplified version of the Freundlich equation (1,2,7) to describe the partitioning or distribution of compounds between the solid and liquid phases.

The Freundlich equation is used to predict the equilibrium
relationship of the concentration of adsorbate in solution

(C) to the mass of adsorbate per unit mass of adsorbent
(x/m), using an expression of the form

$$\frac{x}{m} = K_d C^{(1/n)}$$

where K_d and n are empirical coefficients. In practice,
the partitioning or distribution coefficient K_d is usually
taken to be a constant. The exponent n is usually taken to
be constant and equal to one. The approach presented in
this paper follows these conventions. The value of K_d is
unique for each combination of a specific chemical compound
and a particular soil. It is generally determined by batch
tests or column breakthrough tests in the laboratory. However, in the absence of data from these tests for a particular compound and soil, the value of K_d may be estimated. Such an estimate may be based on published correla-
tions with the compound's solubility in water, with its
biocentration factor, or with its octanol/water partitioning coefficient, together with the organic carbon content
of the soil in the absence of data from these tests for a
particular compound and soil (3,4,5,6).

The Freundlich equation is a simplified description of a
very complex adsorption-desorption process. A more rigorous approach was judged to be inappropriate due to the developmental nature of the science and the uncertainty imposed by assumptions required for site characterization
(e.g., unsaturated and saturated zone parameters, contaminant distribution, and accuracy of laboratory measurements).

APPLICATION OF THE PRINCIPLES

Assumptions

To apply adsorption-desorption principles to a practical
problem at an uncontrolled hazardous waste site, a number
of assumptions must be made:

1. Synergistic effects between compounds are ne-
    glected--each compound is assumed to act indepen-
    dently and in the same manner as it would if it
    were alone in the soil.

2. Compounds remain in contact with the soil long
    enough to reach equilibrium between the contami-
    nants in the water and the contaminants adsorbed
    onto the soil skeleton.

3. The adsorption-desorption relationship is com-
    pletely linear and reversible over the range of
    contaminant concentrations considered.

4.   The partitioning coefficient K_d is a constant for each contaminant.

5.   Compounds move unaltered from their initial location in the source zone to the extraction well, with no biological or chemical degradation.

6.   Measured concentrations of contaminants in the source zone and in the aquifer are accurate. In reality, measurements may vary by 30 percent or more for some compounds and still be considered accurate, depending on the chemical, the analysis method, and the concentration (i.e., how close it is to detection or quantification limits).

Some of these assumptions are probably not particularly close to actual field conditions. Some effects may be additive, while other effects may be compensating. It must be kept in mind, however, that the concepts presented are meant to assist in comparison of alternatives and to serve as an aid to engineering judgment.

First Flush

Consider the cross section shown in Fig. 1. The initial phase of contaminant removal may be thought of as the "first flush," or removal of the contaminated water that is in the aquifer pores at the start of extraction.

Movement of the contaminants is governed by groundwater flow, dispersion, diffusion, and contaminant/aquifer interactions. Accurate prediction of removal of the plume as a whole is difficult, if not impossible. Therefore, another group of reasonable, simplifying assumptions must be made.

As a first approximation, further dispersion and diffusion are neglected, and the plume is assumed to flow toward the extraction wells at a rate dependent on the induced and natural gradients toward the well. The velocity of specific contaminants is reduced by their retardation coefficient. The retardation coefficient is derived from the partitioning coefficient K_d and other physical parameters such as soil bulk density and volumetric water content (1). The velocity of a specific contaminant is then a ratio of the interstitial groundwater velocity divided by the retardation coefficient.

Travel times for the various contaminants are determined based on a range of possible flow path lengths; a range of estimated pore-water velocities, considering both natural and induced groundwater gradients; and retardation coefficients for each compound of interest. Flow path lengths from the extraction well to the leading edge, the peak, and the farthest edge of the plume are estimated. With these, estimates of travel times for contaminant from the peak and plume edges to the extraction wells can be made. Values of the retardation coefficient for each compound are assumed to be constants.

As the first flush occurs, uncontaminated water from upgradient enters the portion of the aquifer influenced by the plume. In this portion of the aquifer, contaminants have adsorbed onto the aquifer skeleton. Upgradient, uncontaminated water entering this area becomes contaminated by desorption of contaminants that had been previously adsorbed onto the aquifer skeleton. In addition, contaminants leaching from the unsaturated zone may be contributing to this portion of the aquifer. Therefore, to estimate long-term variations in groundwater quality (beyond the first flush approach), consideration must be given to contaminants leaching from the unsaturated zone and to possible soil remedial actions that can affect contaminant leaching.

Unsaturated-Saturated Zone Interaction

Water percolates downward through the unsaturated zone at the site and subsequently intersects and combines with the groundwater in the saturated zone. This percolating water leaches or extracts contaminants from the soil and introduces them into the groundwater. Soil remedial actions (such as excavation or capping) influence groundwater quality by affecting two key elements in the leaching process: the quantity of contaminant available for leaching, and the quantity of water percolating through the soil profile.

Groundwater quality directly affects the cost of groundwater remedial actions. Therefore, by considering the relative impact of various soil remedial actions on groundwater quality, an assessment of their relative impact on potential groundwater remedial action cost is possible.

Again, the basis of this analysis relies on the integration of sorption theory with mass balancing. Adsorption-desorption principles are applied to estimate contaminant concentrations in soil and water. Mass balancing is applied to estimate depletion rates for contaminants present in both the unsaturated and saturated zones.

This analysis involves estimating the infiltration rates, the quantity of contaminants available for leaching, the resulting concentrations of those contaminants in water percolating through the unsaturated zone, and the resulting rates of contaminant contribution to the groundwater. It is emphasized that these estimates are tentative and limited in accuracy owing to the many uncertainties involved. They are presented as a practical tool for rationally assessing the relative impacts associated with various remedial actions.

The analysis can be divided into the following major steps:

1. Estimation of Leachate Parameters: Total mass
    and concentration of contaminants in the unsatu-
    rated zone and in leachate versus time.

2. Estimation of Groundwater Parameters: Total mass
    and concentration of contaminants present in the
    saturated zone (on solids and in solution).

3. Estimation of Groundwater Quality Versus Time:
    Combined effects of contaminants present and mov-
    ing in the unsaturated and saturated zones.

Estimation of Leachate Parameters. To estimate contaminant
concentrations in water percolating through the unsaturated
zone beneath the site, the "average" concentration of each
contaminant in the soil is needed. The average soil concentration is determined by dividing the calculated total
mass of each contaminant by the total mass of soil in the
unsaturated zone. The total mass of each contaminant is
estimated based on:

o   Chemical analysis results from soil samples ob-
    tained during onsite investigation

o   Site topography

o   Geology

o   Historical waste storage and handling practices

o   Sample locations

These factors are used to subdivide the soil beneath the
site into blocks of varying volume. In general, each block
is sized to include one soil sample judged to be representative of conditions in the entire block. The mass of each
contaminant in each block is then estimated by multiplying
the mass of soil in the block times the contaminant concentration in the sample. The total mass of each contaminant
in the unsaturated zone can then be determined by totaling
the masses of each contaminant in each of the blocks.

Contaminant concentrations in the leachate are estimated
from these calculated average soil concentrations. The
Freundlich equation is used to describe the adsorption-
desorption process and to estimate the concentrations of
various contaminants in water in the unsaturated zone.

Estimation of the water percolation rate through the unsaturated zone makes it possible to estimate source zone
contaminant mass depletion. Water passing through the unsaturated zone over a period of time removes some of the
contaminants as they desorb from the soil skeleton. This
removal reduces the "average" concentration of each contaminant, thus reducing the calculated concentrations in

the next volume of water passing the profile. Iteration allows an estimation of the changes in contaminant concentration in soil and leachate with time.

Estimation of Groundwater Parameters. To apply a mass balance approach to the groundwater system requires an estimate of the total mass of contaminants present. The mass of each compound in the groundwater is calculated from its measured groundwater concentration and the approximate dimensions of that compound's plume. The Freundlich equation is applied using groundwater contaminant concentrations to estimate the mass of each contaminant adsorbed onto solids in the saturated zone. The sum of the estimated mass of contaminants in the groundwater plus the estimated mass of contaminants adsorbed on the soil skeleton yields the total mass of contaminant present in the saturated zone.

Groundwater Quality Versus Time. Estimation of contaminant concentrations in groundwater with time requires a simulation of the complex processes involved in the mixing zone or interface between the unsaturated and saturated zones. Each compound is considered independently.

Leachate from the unsaturated zone is assumed to enter a "cell" of aquifer volume equal to the volume of the plume for that specific compound. The quantity of leachate that enters the cell is assumed to be equal to the quantity of expected percolation through the soil (based an infiltration rates) during a period of time equal to the estimated "resident time" of groundwater in the cell. The resident time is estimated based on the time it would take groundwater, moving at its porewater velocity, to travel from the up-gradient edge to the down-gradient edge of the plume.

After calculating the volume of leachate that moves into the cell during one resident- time interval, groundwater free of any contaminants (or at upgradient background levels of contamination) is assumed to fill the remaining pore space. The equilibrium concentrations of the groundwater and the aquifer skeleton are determined using the Freundlich equation and the total mass of contaminant available in the cell (sum of the contaminant mass adsorbed on the aquifer skeleton plus that contributed by the entering leachate). The equilibrium groundwater concentration resulting from this analysis is a new "average" concentration within the cell volume. The "equilibrated" groundwater now containing contaminant at the newly calculated concentration is then removed by flow through the cell.

This entire computation process is repeated until sufficient data have been developed to define the concentration versus time curve. Each cycle represents a pore volume exchange and thus a period of time. Each withdrawal of leachate from the unsaturated zone or contaminated groundwater from the aquifer cell volume removes some contaminants, thereby changing the estimated concentrations of leachate and groundwater with time. The total mass of contaminants to be partitioned between the solid and liquid phases thus changes during each cycle owing to mass balance considerations in both the unsaturated and saturated zones.

The authors have developed a program for the IBM PC computer to perform the computations described.

Results

The results of these analyses can be summarized as shown schematically in Fig. 4. For each compound of interest, the contaminant concentration in the groundwater (C_{gw}) as a function of pumping time can be estimated. By estimating the dilution associated with extraction (since some "clean" water will also be captured), the contaminant concentration of the extracted water (C_{ew}) can be estimated with time. Where treatment is a consideration, this curve (for the most critical compound or compounds) will govern treatment system design. A third curve may also be developed that represents the contaminant concentration in the effluent water (C_e). Depending on site conditions and permit requirements, this curve can also incorporate the effect of dilution in the receiving waters. With these curves for each compound of concern, estimates regarding sizing of treatment system, duration of treatment, and duration of pumping can be made.

The size of the treatment system will be predicated on the magnitude of the peak of the C_{ew} curve and the pumping rate. The required duration of treatment can be estimated by determining the time (t_t) at which the discharge criteria concentration (C_{dis}) is no longer exceeded by the C_{ew} curve. Likewise, the required duration of pumping (t_p) can be estimated by determining the time at which the groundwater criteria concentration (C_{crit}) is no longer exceeded by the C_{gw} curve.

These curves can be developed for all compounds of interest. Certain compounds, because of their mobility characteristics and/or treatability characteristics, may govern extraction and treatment design. Curves for other compounds may demonstrate the futility of attempting "complete" cleanup.

The effectiveness of soil excavation and/or capping in various combinations can also be assessed with this approach. This is done by repeating the analysis, varying the assumed depth of excavation and the infiltration rate. These assumptions will reduce both the "average" soil concentration in the source zone and the leachate migration rate. The results can be used to compare costs for various combinations.

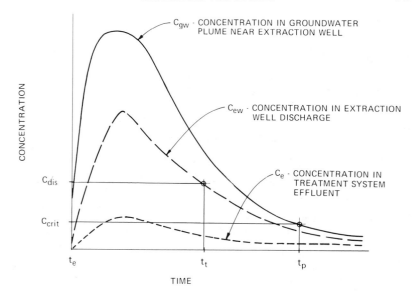

FIG. 4 — Concentration Variation with Time for a Single Contaminant

CONCLUSIONS

The rate of contaminant migration from an uncontrolled hazardous waste site is difficult to predict. By making a number of assumptions, it is possible to use simple equations of sorption theory to estimate groundwater and soil skeleton contaminant concentrations. By using iteration, mass balancing, the Freundlich equation, and estimates of the initial soil and groundwater contaminant concentrations, the organic contaminant concentration in extracted water over a period of time can be approximated. The relative effects of excavation and/or capping of contaminated soil can also be estimated using the same procedures. The procedure presented is believed to be a simple but useful tool for comparing potential remedial actions at uncontrolled hazardous waste sites.

ACKNOWLEDGMENTS

The writers are grateful for the support of CH2M HILL in the preparation of figures and typing for this paper.

Critical review of the manuscript by Lawrence H. Roth is sincerely appreciated.

APPENDIX I.--REFERENCES

1. Freeze, R.A., and Cherry, J.A., Groundwater, Prentice-Hall, New Jersey, 1979.

2. Freundlich, H., Colloid and Capillary Chemistry, translated from the 3rd German edition by H. Stafford Hatfield, Methuen & Co., Ltd., London, 1926.

3. Karickhoff, Samuel W., Brown, David S., Scott, Trudy A., "Sorption of Hydrophobic Pollutants on Natural Sediments," Water Research, Vol. 13, 1979, pp. 241-248.

4. Kenaga, Eugene E., "Predicted Bioconcentration Factors and Soil Sorption Coefficients of Pesticides and Other Chemicals," Ecotoxicology and Environmental Safety, Vol. 4, 1980, pp. 26-38.

5. Kenaga, E.E., and Goring, C.A.I., "Relationship Between Water Solubility, Soil Sorption, Octanol-Water Partitioning, and Bioconcentrations of Chemicals in Biota," Aquatic Toxicology, ASTM STP 707, J.C. Easton et al., eds., American Society for Testing and Materials, Philadelphia, Pennsylvania, 1980.

6. Lyman, Warren J., Reehl, William F., and Rosenblatt, David H., Handbook of Chemical Property Estimate Methods, McGraw-Hill, Inc., New York, 1982.

7. Bohn, Hinrich, Soil Chemistry, John Wiley & Sons, Inc., New York, 1979.

APPENDIX II.--NOTATION

The following symbols are used in this paper:

C = Concentration of adsorbate in solution;

C_e = Concentration of contaminant in treatment system effluent;

C_{ew} = Concentration of contaminant in well discharge;

C_{gw} = Concentration of contaminant in groundwater;

C_s = Concentration of contaminant in soil;

K_d = Partitioning coefficient;

m = Mass of contaminant in soil;

n = Exponent in partitioning equation (normally taken as 1);

t_e = Time at which extraction pumping begins;

t_p = Time at which pumping may be stopped;

t_t = Time at which treatment may be stopped;

x = Mass of adsorbate.

WATER BALANCE APPROACH TO PREDICTION OF
SEEPAGE FROM MINE TAILINGS IMPOUNDMENTS

PART I

GENERAL WATER BALANCE APPROACH AND SOME TYPICAL MODELING RESULTS

Keith A. Ferguson[1], A.M. ASCE, Ian P.G. Hutchison[2],
Robert L. Schiffman[3], M. ASCE

Abstract

The prediction of seepage from a tailings impoundment requires careful evaluation because the predictions can have significant economic impact upon designs, as well as permit stipulations and long-term seepage treatment and monitoring requirements. A general methodology called the "Water Balance" approach is presented which can be used to predict the quantity and quality of seepage of interstitial tailings water from mine tailings impoundments. The water balance approach is used to account for the most important mechanisms causing seepage including seasonal and annual variations in climate, and surface and groundwater hydrology, as well as the consolidation of the tailings. A detailed discussion of Finite Strain Consolidation theory as applied to the consolidation aspect is given. A computer program developed for water balance analyses as well as results from a case history are also discussed.

Introduction

Seepage losses from tailings impoundments are traditionally determined by using methodologies such as flow nets, or analytic or numerical solutions to equations describing steady state flow through porous media. These methods assume that water supply is unlimited and that the seepage losses are a function of relatively constant driving head(s) and hydraulic properties of the media through which the seepage flows. While this approach is valid for water dams, and may be suitably conservative for some types of waste impoundments, it may not be directly applicable or appropriate for tailings disposal facilities. For these types of problems, an approach which involves the determination of the temporal variation in the volume of water stored in various portions of the impoundment is required. This paper describes this type of "Water Balance" approach.

[1]Project Engineer, Steffen Robertson and Kirsten (Colorado) Inc.
[2]Division Head-Water Engineering, Steffen Robertson and Kirsten (Colorado) Inc.
[3]Professor of Civil Engineering, University of Colorado, Boulder.

The key factors influencing seepage from a tailings impoundment include:

1) The site's surface and groundwater hydrologic regime;
2) The depositional pattern and behavior of tailings;
3) The consolidation characteristics of the tailings, and the manner in which these are influenced by the depositional environment and sequence;
4) Long-term interstitial tailings water discharge as a result of reclamation activities, gravity drainage, and secondary consolidation; and
5) Long-term seepage as a result of surface water infiltration.

The paper is organized in two parts. Part I describes the overall water balance approach including an identification of all components and a discussion of the analytical methods used to predict the behavior of these components. Part I includes a discussion of a computer program which has been developed to simulate a tailings impoundment water balance over an extended period of time and presents some summary results of computer modeling studies completed for case history. Part II presents a detailed discussion of the key theoretical aspects associated with some of the water balance components. Detailed results from a case history are presented in Part II to demonstrate the relative magnitude of expected seepage from the various water balance components.

Complex problems such as the one described in this paper are more frequently requiring multidisciplinary efforts involving surface and groundwater hydrologists, geotechnical engineers, and hydrogeochemists. It is hoped that the approach described in this paper will increase awareness of the need for interdisciplinary efforts, and that future problems solved with the water balance approach will contribute to improved management of tailings impoundments.

General

The water balance approach to seepage modeling requires the definition of a boundary around the tailings impoundment within which inflow and outflow components can be determined and balanced over a specified time increment. Appropriate time increments can be selected in order to evaluate or incorporate time dependent variations in surface or groundwater hydrology, material properties of the tailings, depositional history, or other important factors affecting seepage quantities and quality. To illustrate this, a typical cross-section of an unlined cross-valley impoundment during operation is shown in Figure 1. The water balance components are indicated by the arrows and are summarized in Table 1. The overall impoundment water balance has been subdivided into the water balance of the surface pond, the reservoir formed by permeable soil and rock located below the tailings, and the downstream embankment shell. The water balance approach involves tracking the mass of water (and possibly key chemical constituents) in each of the three storage systems. The resulting quantity (and quality) of outflow from the downstream embankment shell represent the seepage escape from the impoundment.

A cross-valley impoundment is typical of the depositional environments of a high percentage of tailings impoundments in the western United States. Other types of waste disposal sites such as side-hill, or ring dike systems are also frequently designed and built, and can be characterized in a similar fashion to the cross-valley impoundment.

A further, more detailed breakdown of some of the components is required in order to facilitate the analysis. For example, tailings slurried to the impoundment at a specified density includes a volume of solids (V_s) and a volume of water (V_w) per unit time (T). Once in the impoundment, the volume of water must be evaluated as several separate subcomponents, that is:

V_{sd} = Volume of water which is discharged to the impoundment pool as result of sedimentation processes;

V_{cu} = Volume of water which drains to the upper tailings surface during primary consolidation of time increment T

V_{cd} = Volume of water which drains to the lower tailings surface during primary consolidation of time increment T

V_{vs} = Volume of water seeping into embankment from the tailings during primary consolidation

V_i = Volume of water retained in tailings (interstitial water)

V_{misc} = Volume of water seeping into the impoundment foundation or embankment as a result of gravity drainage or other miscellaneous processes.

$\sum$ = V_w

If the water quality of the various water sources is known, then the mass balance modeling can be extended to include the mass of certain key conservative constituents.

Following closure and reclamation, a slightly different set of balance components requires consideration. These are illustrated on Figure 2. The geotechnical considerations become much more significant during this period due primarily to gravity drainage, secondary consolidation, and surface infiltration aspects. A summary of the water balance components considered during this phase of the impoundment life are given in Table 2. It is important to reiterate that the inflow and outflow components given in Tables 1 and 2 are in most cases time dependent. Different components will have differing degrees of "activity" at any given time. The critical importance of the water balance approach is consideration of the "activity" of the various components at any time (or over any time increment) to get $\sum A = \sum B$.

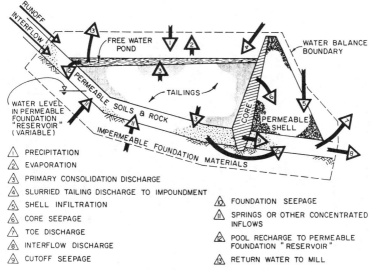

- ⚠1 PRECIPITATION
- ⚠2 EVAPORATION
- ⚠3 PRIMARY CONSOLIDATION DISCHARGE
- ⚠4 SLURRIED TAILING DISCHARGE TO IMPOUNDMENT
- ⚠5 SHELL INFILTRATION
- ⚠6 CORE SEEPAGE
- ⚠7 TOE DISCHARGE
- ⚠8 INTERFLOW DISCHARGE
- ⚠9 CUTOFF SEEPAGE
- ⚠10 FOUNDATION SEEPAGE
- ⚠11 SPRINGS OR OTHER CONCENTRATED INFLOWS
- ⚠12 POOL RECHARGE TO PERMEABLE FOUNDATION "RESERVOIR"
- ⚠13 RETURN WATER TO MILL

FIGURE I. TYPICAL WATER BALANCE COMPONENTS
DURING IMPOUNDMENT OPERATION (UNLINED
IMPOUNDMENT.)

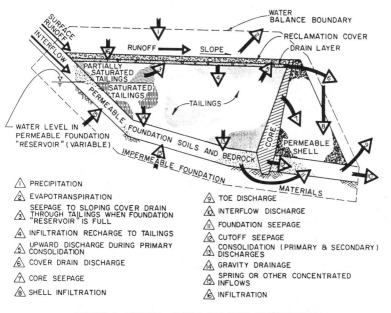

- ⚠1 PRECIPITATION
- ⚠2 EVAPOTRANSPIRATION
- ⚠3 SEEPAGE TO SLOPING COVER DRAIN THROUGH TAILINGS WHEN FOUNDATION "RESERVOIR" IS FULL
- ⚠4 INFILTRATION RECHARGE TO TAILINGS
- ⚠5 UPWARD DISCHARGE DURING PRIMARY CONSOLIDATION
- ⚠6 COVER DRAIN DISCHARGE
- ⚠7 CORE SEEPAGE
- ⚠8 SHELL INFILTRATION
- ⚠9 TOE DISCHARGE
- ⚠10 INTERFLOW DISCHARGE
- ⚠11 FOUNDATION SEEPAGE
- ⚠12 CUTOFF SEEPAGE
- ⚠13 CONSOLIDATION (PRIMARY & SECONDARY) DISCHARGES
- ⚠14 GRAVITY DRAINAGE
- ⚠15 SPRING OR OTHER CONCENTRATED INFLOWS
- ⚠16 INFILTRATION

FIGURE 2. TYPICAL WATER BALANCE COMPONENTS
FOLLOWING RECLAMATION (UNLINED IMPOUNDMENT)

TABLE 1
SUMMARY OF WATER BALANCE COMPONENTS
DURING IMPOUNDMENT OPERATION

Freewater Pond:

INFLOWS (A)	OUTFLOWS (B)
1) Water from settling of tailings	1) Return flow to mill
2) Precipitation	2) Evaporation
3) Primary consolidation discharge	3) Foundation "Reservoir" discharge
4) Runoff	4) Core seepage
5) Foundation "Reservoir" overflow	

$$\sum (A) = \sum (B) \quad \text{For water balance}$$

Permeable Foundation Reservoir:

INFLOWS (A)	OUTFLOWS (B)
1) Consolidation discharge	1) Cutoff seepage
2) Interflows	2) Foundation discharge
3) Springs or other concentrated inflows	3) "Reservoir" overflow to freewater pond
4) Freewater pond overflows	4) Saturation water (reservoir filling)

$$\sum (A) = \sum (B) \quad \text{For water balance}$$

Toe and Interflow Discharge:

INFLOWS (A)	OUTFLOWS (B)
1) Core seepage	1) Toe seepage
2) Cutoff seepage	2) Interflow discharge
3) Foundation seepage	3) Saturation waters (to sat. or elevate the moisture content of previously partially sat. materials)
4) Shell infiltration	

$$\sum (A) = \sum (B)$$

TABLE 2
SUMMARY OF WATER BALANCE COMPONENTS
AFTER RECLAMATION

Permeable Foundations Reservoir:

INFLOWS (A)	OUTFLOWS (B)
1) Consolidation discharge	1) Cutoff seepage
2) Gravity drainage	2) Foundation seepage
3) Interflow	3) Foundation "Reservoir" overflow to cover drain
4) Springs or other concentrated inflows	4) Seepage to sloping cover drain through tailings when foundation "Reservoir" is full
5) Infiltration through reclamation cover	5) Core seepage component when foundation "Reservoir" is full
	6) Saturation water (reservoir filling)

$$\sum (A) = \sum (B)$$

Cover Drain Discharge:

INFLOWS (A)	OUTFLOWS (B)
1) Cover infiltration	1) Infiltration recharge to tailings (when tailings are unsaturated)
2) Upward discharge from tailings during consolidation	2) Cover drain discharge
3) Foundation "Reservoir" overflow to cover when full	
4) Seepage to sloping cover drain through tailings when foundation "Reservoir" is full	

$$\sum (A) = \sum (B)$$

Toe and Interflow Discharge:

INFLOWS (A)	OUTFLOWS (B)
1) Cutoff seepage	1) Toe discharge
2) Core seepage	2) Interflow discharge
3) Foundation seepage	3) Saturation waters
4) Infiltration	

$$\sum (A) = \sum (B)$$

The following sections give a brief description of the major components of the water balance model.

Components of Water Balance Models

(a) Surface Hydrologic Components

The two major components include surface runoff directly into the free water pond and interflow, or near-surface groundwater flow which enters the permeable soil and rock zone immediately below the tailings. The surface runoff component can be estimated by using the drainage area of the catchment surrounding the tailings impoundment and the unit runoff values derived from gaged data in the vicinity. Interflow is a more difficult component to estimate. Approximate values can be derived from a study of measured hydrographs from the catchment or a similar catchment area in the vicinity or by estimating the depth and permeability and, hence, flow capacity of the surficial material. It is usually appropriate to express interflow as a fraction of the runoff.

Computer models are available that will simulate the runoff and interflow response of a catchment subject to a given precipitation and potential evaporation sequence. Computer models are appropriate where impoundments are constructed in wet and highly variable climates and surface runoff and interflows have a significant effect upon the overall water balance evaluations.

Both runoff and interflow vary on a seasonal and annual basis. It is, therefore, necessary to obtain a relatively long sequence of monthly values in order to determine average seasonal fluctuations and the range of the yearly totals. Water quality of these components is generally well defined. This water usually does not contain any constituents at significant concentrations.

(b) Groundwater Flow Component

The groundwater flow component can include localized springs or a more general seepage of groundwater into the permeable zone below the tailings. These would be generally more important in wetter climates and near the bottom of valleys where groundwater outflow conditions occur. These seepage rates fluctuate seasonally and annually but usually to a lesser extent than the surface runoff and interflow components. Estimates of the magnitude of these components can be derived from a knowledge of the regional groundwater flow regime and spring flow measurements taken prior to placement of the tailings.

It is important to note that springs or other concentrated inflows may become locations of concentrated outflow under sufficiently high heads that can be produced by the development of the impoundment. This factor should be considered when evaluating groundwater flow components.

Water quality considerations may be important where these components are concerned. Springs or other concentrated inflows may represent acid drainage with elevated dissolved trace metal contents. In such cases, they can have a significant effect upon the quality of the seepage escaping from the impoundment. Field data is required to adequately define the concentrations of the important trace metals in such springs.

(c) Geotechnical Components

The geotechnical aspects requiring evaluation can be generalized into the following five components:

· Core Seepage;
· Foundation Seepage;
· Consolidation Discharge;
· Gravity Drainage; and
· Surface Infiltration.

The discussion below describes the most significant considerations and factors which need to be included in water balance analyses. The theoretical basis of some of the analysis methods are given in Part II.

(i)   Core Seepage

The availability of water for core seepage is a combined source of free water on the impoundment surface and consolidation discharge below the free-water surface. Tailings deposited in the impoundment have a wide range of void ratios (ratio of the volume of voids to the volume of solids). As a result, the permeability of the tailings at the contact with the upstream slope of the dam is highly variable. This is illustrated on Figure 3 which shows permeability as a function of void ratio for some typical tailings materials. Other pertinent data related to the five samples of material shown on Figure 3 are given in Table 3.

TABLE 3
SUMMARY OF PHYSICAL CHARACTERISICS OF
FIVE SAMPLES OF TAILINGS SHOWN IN FIGURE 3

Sample	#Description	% Passing #200 Sieve	Specific Gravity	Atterberg Limits	
				Liquid Limit	Plasticity Index
Sample 1	Sulfide Tailing	95	4.2	-	-
Sample 2	Gold Tailing	90	2.8	-	NP*
Sample 3	Gold Tailing	94	3.1	34.8	1.2
Sample 4	Gossan Tailing	92	3.0	-	-
Sample 5	Gold Tailing	80-89	3.0	-	NP

* NP = Non-plastic

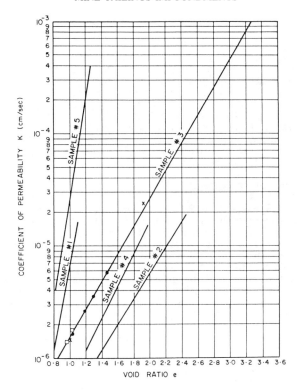

FIGURE 3. SOME TYPICAL RELATIONSHIPS BETWEEN VOID
RATIO AND COEFFICIENT OF PERMEABILITY FOR
5 DIFFERENT MINE TAILINGS MATERIALS

The rate of rise in a typical tailings impoundment
located in mountainous terrain may be as much as 50 to 100
feet during the first year or two of operation. The
typical range of effective stresses in the tailings during
this period can result in void ratios ranging from over 3
to less than 1.5. The results in Figure 3 indicate that a
possible variation of two to three orders of magnitude in
tailings permeability may exist along the upstream face of
the dam.

A further important factor affecting core seepage is
the rate of consolidation during operation and immediately
following closure. An added complication is consolidation
as a result of reclamation activities. Mittal and
Morgenstern (1976) have demonstrated that conventional
steady-state analysis techniques are not appropriate for
under-consolidated tailings. Conventional flow net
evaluations can result in erroneous predictions of core

seepage during the primary consolidation phase. For
tailings impoundments where materials are typically
retained by a full-height earth structure with a cutoff
system, an important consideration is the relative
permeability of the core with respect to the water
available for discharge from the primary consolidation
process. Following completion of primary consolidation,
flow nets can be used to give a reasonable prediction of
seepage provided the effects of tailings permeability,
variable driving heads and specific retention (field
capacity) of the tailings are considered. The last two
considerations are important if the saturation level in
the tailings decreases with time.

(ii)   Foundation Seepage

        Seepage through the impoundment foundation requires an
understanding of the driving hydrostatic head and the
hydraulic parameters of the foundation and embankment
materials. This driving head is in turn, a function of
permeable foundation storage capacity and the volume of
water entering this storage. The most significant inflow
components include groundwater, surface water recharge,
consolidation discharge from the lower tailings surface
and gravity drainage. By using conventional flow net or
analytical techniques, foundation seepage can be estimated
from the driving head available in the permeable upstream
foundation reservoir. The rate of foundation seepage can
then be determined through the water balance technique by
an interactive analysis of balancing inflows, outflows,
and net change in storage in the foundation reservoir
area.

(iii)   Consolidation Discharge

        The prediction of interstitial water discharge from
the tailings during primary consolidation is most easily
accomplished through the use of advanced consolidation
theories and numerical modeling techniques. Correct
modeling requires consideration of the rate of rise of the
impoundment, and boundary conditions at the base and
surface of the impoundment for operation and post-closure
site conditions. Finite strain consolidation theory is the
best available technology for these predictions (See Part
II).

        Numerical solutions and available computer programs
currently limit the type of detailed consolidation
evaluation to that of a one-dimensional analysis. A cross-
valley type of impoundment typically imposes complex
geometrical considerations upon the interpretation of one-
dimensional analysis results. The geometrical effects of
other types of impoundments are generally less

significant. The current state of practice allows the geometrical effects to be addressed in a qualitative fashion
only. Additional research and development of computer
models to evaluate complex boundary and geometrical
effects will be necessary to determine quantitative data
for a water balance evaluations. The quantitative data
presented in Part II illustrates the fundamental
difference between one-dimensional analyses methods.

(iv) Gravity Drainage

        Gravity drainage is controlled by several factors, the
most significant of which are boundary conditions. A "dry"
condition at the base of the impoundment will allow
gravity drainage to begin even during the primary consolidation process. A saturated base condition will limit
gravity drainage until consolidation excess porewater
pressures have dissipated. Once excess porewater pressures
have dissipated, gravity drainage begins only when the
water level in the foundation reservoir falls below the
level in the tailings or as impoundment seepage occurs.

        The other significant factors affecting the rate of
gravity drainage include the permeabilty characteristics
of the tailings and impoundment geometry. The total volume
of water available for gravity discharge depends most
significantly upon the specific retention of the tailings.
This parameter, sometimes referred to as the field
capacity, is the water which is intrinsically retained in
the tailings interstices against the forces of gravity.
The results from finite strain consolidation analyses can
be evaluated to determine spatial variation of the
saturated tailings permeability, which in turn can be used
to aid in the analysis of gravity drainage.

(v)   Surface Infiltration

        Surface infiltration begins to occur following the
cessation of water discharge to the impoundment surface
due to primary consolidation in the tailings impoundment.
The volume and rate of surface infiltration is most
significantly affected by climatic conditions, the perm-
meability characteristics of the impoundment reclamation
surface, impoundment geometry, and the overall rate of
impoundment seepage and the resulting drop in the saturation level in the tailings. Surface infiltration requires
consideration of the evapotranspiration processes and partially saturated conditions throughout the infiltration
profile as well as in the upper portions of the tailings.

Water Balance Computer Modeling

        The prediction of seepage from a tailings impoundment using a
water balance approach requires tracking a large number of time-

varying water volume and chemical mass components. Computer programs are the easiest means of dealing with this.

A schematic diagram representing the tailings impoundments shown in Figures 1 and 2, is illustrated in Figure 4. This schematic diagram is the basis of a computer program developed for analyzing a tailings impoundment water balance. A simplified chemical mass balance routine has also been incorporated into the model for purposes of tracking the concentration of certain key chemical constituents.

The computer program analyzes the water balance for specified time increments such as weekly or monthly. Overall time frames of up to 50 or more years can be handled. The program currently requires input data for items such as consolidation discharge, percolation through the reclamation profile, core, cutoff, and foundation seepage, and tailings recharge. The water flow input data is generated through separate analyses by using other computer programs or simplified techniques. Chemical constituent concentrations are obtained from sampling of the various types of water.

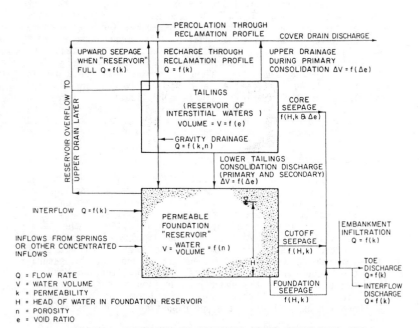

FIGURE 4: SCHEMATIC OF COMPUTER MODEL DEVELOPED FOR OVERALL WATER BALANCE EVALUATIONS & PREDICTION OF IMPOUNDMENT SEEPAGE FOLLOWING RECLAMATION

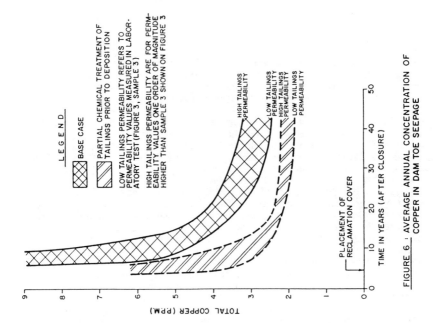

FIGURE 6 : AVERAGE ANNUAL CONCENTRATION OF COPPER IN DAM TOE SEEPAGE

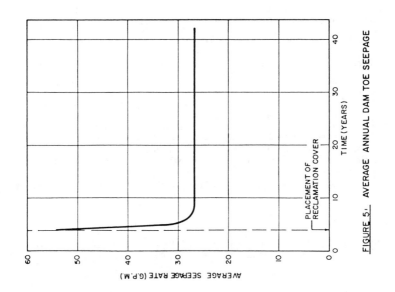

FIGURE 5 : AVERAGE ANNUAL DAM TOE SEEPAGE

Typical Model Results

The model described was applied to a case involving a tailings impoundment in a relatively wet climate. It was important to estimate the long-term quantity and quality of seepage emanating from the toe of the dam. This information was required by the operators to determine the extent and duration of a seepage treatment facility that was required so that local water quality standards could be met.

The model was set up as described above and used to simulate the seepage rate and quality for a period of 45 years after closure. Sensitivity runs were conducted to determine the influence of different levels of pre-treatment of the tailings before deposition in the impoundment upon seepage water quality.

Input to the model included primary and secondary consolidation discharge volumes as well as surface infiltration. Gravity drainage was evaluated as part of the overall foundation reservoir water balance. An order of magnitude variation of tailings permeability was also considered in the sensitivity runs for evaluation of discharge water quality. Some of the model output is illustrated in Figures 5 and 6. Figure 5 shows the average annual seepage rate. The long term seepage rates were based upon actual field measurements of seepage quanities. Because data was available on seepage for a wide range of driving heads in the permeable foundation reservoir, sensitivity analyses were not required. If water balance evaluations are completed as part of a design phase, sensitivity analyses for a range of foundation hydraulic parameters should be considered. Figure 6 shows the corresponding estimated copper concentrations in the seepage water. The model was run using a monthly time step in order to look at the seasonal variation in these values.

Conclusion

Because of the inherent difficulty associated with predicting permeability of natural foundations and modeling the complex hydrochemical behavior in a tailings impoundment, the model in its current form can be regarded as providing order-of-magnitude estimates of seepage and the flow water quality. It provides a sys-tematic representation of the processes that occur and an indication of their effect on the seepage water quality. In general, tailings consolidation and gravity drainage flows are the major contributions to high levels of contaminants in seepage exiting the embankment toe during the first few years following closure; thereafter, core seepage, gravity drainage, and infiltration become the most important factors.

References

Mittal, H. K., and Morgenstern, N.R., (1976), "Seepage Control in Tailings Dams", Canadian Geotechnical Journal, 13, No. 3, pp. 277-293.

WATER BALANCE APPOACH TO PREDICTION OF
SEEPAGE FROM MINE TAILINGS IMPOUNDMENTS

Part II

THEORETICAL ASPECTS OF WATER BALANCE
APPROACH TO SEEPAGE MODELING AND DETAILED CASE HISTORY RESULTS

Keith A. Ferguson[1], A.M. ASCE, Ian P.G. Hutchison[2],
Robert L. Schiffman[3], M. ASCE

General

The prediction of seepage from a mine tailings impoundment requires careful evaluation because the predictions can have significant economic impact upon designs, as well as permit stipulations and long-term monitoring requirements. The need for careful evaluation is compounded if the seepage contains contaminants. The primary consolidation of tailings occurs over a relatively short term. The other items such as secondary consolidation, gravity drainage, and infiltration occur over much longer periods of time and, therefore, control long-term seepage rates and contaminant migration from a tailings impoundment.

Some of the most significant volumes and rates of interstitial tailings water discharge generally occurs during the initial portions of the primary consolidation process. During primary consolidation, the rate of water discharge typically is not constant and volumes of discharge are a direct function of the overall settlement magnitudes. The theoretical considerations of the primary consolidation processes as related specifically to the prediction of interstitial water discharge rates, volumes, and engineering characteristics of the tailings affecting long-term seepage rates are discussed in the following section.

Theoretical Considerations - Primary Consolidation

In Part I it was noted that the characterization of seepage from a tailings impoundment requires an accurate account of the rate and volumes of interstitial water discharge. An important aspect in evaluating these parameters is the selection of a consolidation theory and numerical analysis method (Caldwell, et., al, 1984).

[1]Project Engineer, Steffen Robertson and Kirsten (Colorado) Inc.
[2]Division Head-Water Engineering, Steffen Robertson and Kirsten (Colorado) Inc.
[3]Professor of Civil Engineering, University of Colorado, Boulder.

The publication of the theory of consolidation by Terzaghi (1923) was the founding moment for modern geotechnical engineering and since that time, significant advances in theoretical developments, numerical analysis methods, and hands-on experience related to this fundamental problem have occurred. Two principal schools of approach are active today. These include infinitesimal strain theories and finite strain theory. Summaries of these theories are given below:

Infinitesimal Strain Theories:

1)	Classical one-dimensional theory	Terzaghi and Fröhlich (1936)
2)	Uncoupled analysis theory (3-dimensional)	Terzaghi and Rendulic (Schiffman, et al, 1969)
3)	Coupled analysis theory (3-dimensional)	Biot (1941, 1956)

Finite Strain Theory:

1)	One-dimensional theory with non-linear soil properties	Gibson, England and Hussey (1967) Mikasa (1963, 1981)

The basic formulations and governing equations of these theories are well documented and will not be presented here. The recent work by Schiffman, et al (1984) and Croce, et al (1984), have demonstrated through the results of limited field testing and centrifugal studies that finite strain theory is the best available technology for the evaluation of mine tailings settlement rates and magnitudes. Work by Gibson, et al (1967) has shown that one significant inconsonant aspect between Terzaghi-Fröhlich theory and finite strain theory is the prediction of the pore fluid velocity, v_f. In classical theory, v_f (average velocity) is given by the following relation:

$$v_f = - \frac{c_v}{e(1 + e)^2} \times \frac{\delta e}{\delta d} \tag{1}$$

whereas v_f (average velocity) according to the finite strain theory is given by:

$$v_f = - \frac{c_f}{e(1 + e_0)} \times \frac{\delta e}{\delta d} \tag{2}$$

linear theory;

$$c_f = - \frac{(1 + e_0)}{(1 + e)} c_v$$

therefore;

$$v_f = - \frac{c_v}{e(1 + e)} \times \frac{\delta e}{\delta d} \tag{2a}$$

non-linear theory;

$$c_f = - \frac{(1 + e_0)^2}{(1 + e)^2} c_v$$

therefore;

$$v_f = - c_v \frac{(1 + e_0)}{e(1 + e)} \times \frac{\delta e}{\delta d} \tag{2b}$$

where: e   =   void ratio (average over interval d)

        e_0 =   Lagrangian void ratio or void ratio at time t = 0

        c_v =   coefficient of consolidation

        d   =   depth interval

As can be seen in the above relations, the velocity of the pore fluid according to the linear finite strain theory (Gibson et al, 1967) is a factor of (1 + e) faster than that predicted by the Terzaghi theory. The non-linear finite strain theory predicts a pore fluid velocity which is a factor of (1 + e_0) faster than the Terzaghi theory. These are subsequently referred to as difference factors.

The typical range of void ratios in mine tailings impoundments may range from an e_0 of 5 to a final e after self-weight consolidation of 1 to 1.5. If we take the case of e_0 = 3 and at time t where e = 2, the factors above actually correspond to velocities which are three and four times faster (linear and non-linear finite strain theory, respectively) than the predictions from conventional (Terzaghi) theory. Since the volume of flow is directly related to the velocity of flow by an area factor which is consistant to any applied theory, we see that the volume of flow can vary from six (e_0 = 5, e = 5) to two times greater (e = 1 condition using linear finite strain theory) using the finite strain theories, when compared to the volume rate of flow which would be predicted by Terzaghi theory.

The significance of the above differences is best illustrated by example. Using the void ratio versus permeability relationship for Sample 3 given in Figure 3 (Part I) and the void ratio versus effective stress relationship given in Figure 1 (actual gold tailing results) as well as conventional c_v results for an average void ratio of e = 1.6, comparative analyses were completed. The analyses were for the case of a normally consolidated tailings subjected to a uniform surcharge loading of 800 psf during reclamation activities. The following assumptions were made:

1) The tailings were assumed to be normally consolidated over a range of depths in the impoundment from 0 to 120 ft;
2) Saturated conditions exist in all areas of the impoundment;
3) Hydrostatic, free-draining boundary conditions existed at the base of the impoundment; and
4) The impoundment surface was free-draining.

A constant c_v conventional analyses was used. The c_v values were taken from the test results shown in Figure 2. A c_v = 1.2 x 10^{-1} cm^2/sec, at e = 1.6 was selected on the basis of the predicted void ratio distribution from the finite strain results. Although a typical conventional analysis would not be privileged to this data, it was used for these comparative analysis in order to illustrate fundamental theoretical differences.

The results of the comparative analyses are given in Figures 3 through 5. Figure 3 shows the predicted degree of consolidation versus time for the two theoretical approaches. Similarly, Figure 4 shows total settlement versus time and Figure 5 shows a normalized settlement or discharge rate versus time for the two theories. As can be seen on Figure 4, the two theories predict significantly different total settlements. Finite strain theory predicts higher total settlements, it predicts a much higher settlement rate. The results also show another significant difference between the two analyses methods. The times to a 40-percent degree of consolidation in the 120-foot profile (Figure 2) are ten days (finite strain), and 200 days (conventional theory). The difference is a predicted factor of 20. For the 30-foot profile, the times to a 40-percent degree of consolidation are three (finite strain), and ten days (conventional theory). This difference corresponds to a predicted factor of 3.3. Results indicate that larger profiles show an increasing degree of deviation from the conventional theory predictions.

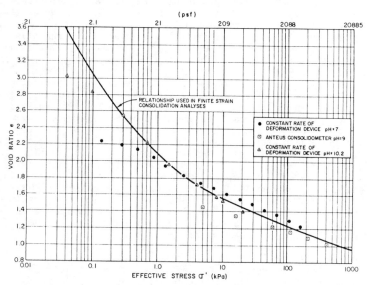

FIGURE I ' VOID RATIO vs. EFFECTIVE STRESS RELATIONSHIP
FOR COMPARATIVE ANALYSIS TAILINGS

The results in Figure 5 show a similar trend. The predicted rates of discharge for the 30-foot profile are consistant with the v_f difference factors previously discussed. The average initial void ratio in the 30-foot profile prior to surcharge loading was approximately 1.5 to 1.7. The $(1 + e_0)$ factor is, therefore, approximately 2.5 to 2.7. At day 1 following loading, the finite strain theory predicts a volume rate of discharge of 0.50 ft/day and conventional theory predicts a volume rate of discharge of 0.20 ft/day. This corresponds to a difference factor of 2.5. The predicted rates for the 120-foot profile at day 1 are 1.0 ft/day (finite strain), and 0.10 ft/day (conventional theory). These numbers correspond to a difference factor of ten which is considerably higher than the factor of 2.5 for the 30-foot profile. The one day values assume an instantaneous placement of the reclamation profile. Although this is not a practical loading rate, it can be used to illustrate fundamental theoretical considerations.

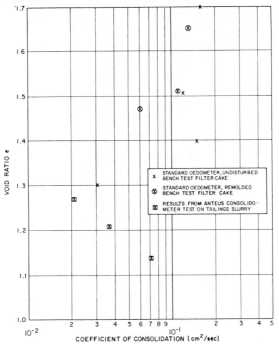

FIGURE 2 : COEFFICIENT OF CONSOLIDATION vs. VOID
RATIO FOR COMPARATIVE ANALYSIS
TAILINGS

These results indicate another important aspect of the theories of consolidation. The v_f relationships previously discussed are relationships for thin layer systems. They reflect minimum difference factors which neglect the effects of self-weight during the consolidation process. The thicker the tailings are in an impoundment, the more significant the self-weight aspect, and subsequently, the difference factor may increase to ten or more.

Discharge from the tailings occurs during consolidation at free-draining boundaries. During self-weight consolidation, the surface of the tailings will always be a drainage boundary. Drainage at the base of the impoundment can generally be modeled in one of three ways (Schiffman, et. al., 1984).

1) Impervious;
2) Free draining under hydrostatic conditions; and
3) "Dry" permeable boundary.

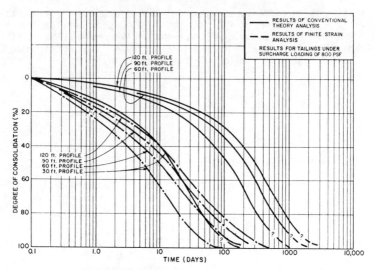

FIGURE 3: PRIMARY CONSOLIDATION RESULTS;
DEGREE OF CONSOLIDATION vs. TIME

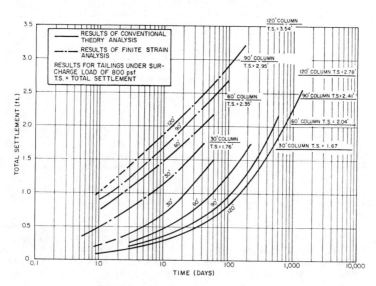

FIGURE 4: PRIMARY CONSOLIDATION RESULTS; TOTAL
SETTLEMENT vs. TIME

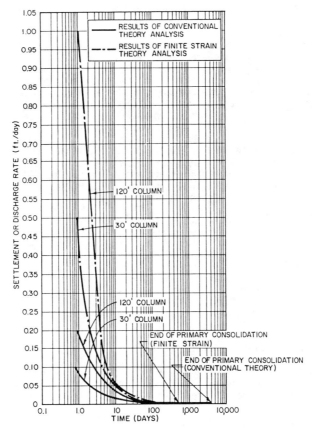

FIGURE 5 : PRIMARY CONSOLIDATION RESULTS;
SETTLEMENT DISCHARGE RATE vs. TIME

Free draining under hydrostatic conditions implies that water
buildup occurs in the foundation as a result of impeded flow (for
example from a cutoff constructed beneath the tailings dam embankment). The "dry" boundary is the condition that is present when the
discharge rate from the tailings is slower than the rate at which the
foundation can transport the water.

Following reclamation, a similar definition of boundary
conditions is required. Of particular concern is the type of drainage
which occurs at the surface of the tailings due to construction of
the reclamation profile. Impedance of discharge water at this surface
can result in impeded consolidation rates. The design of any drainage
system at the tailings surface must be capable of transmitting the
large discharge rates which occur at the initial stages of consolida-

tion. Drainage systems designed on the basis of conventional theory may impede the consolidation process.

Pore pressure isochrones can be utilized to predict the volume of water which discharges to the surface and the base of the impoundment. Normalized isochrones for the 30-foot profile previously discussed are shown in Figure 6. The isochrones shown in Figure 6A are for the 30-foot profile under a surcharge loading of 800 psf with a free-draining surface and a free-draining base under hydrostatic conditions. The isochrones represent the total excess pore water pressure (hydrostatic plus consolidation excess pore water pressures). Flow occurs as a result of the gradients indicated by the slope of the total pore water pressure isochrones. The neutral axis is the line which represents a flow gradient of zero. The volume change above this axis represents water discharging to the surface of the impoundment, and below the neutral axis represents water discharging to the impoundment foundation.

The effect of assigned boundary conditions on the seepage to the upper and lower drainage surfaces can be seen by comparing the relative differences between the total and excess pore-water pressure isochrones in Figure 6B. The isochrones shown in this figure represent porewater pressures during consolidation for a saturated, free-draining base boundary condition. Excess pore water pressure isochrones, would be the total pore water pressure isochrones for a free-draining "dry" type of base boundary condition. It is important to note that the excess porewater pressure isochrones shown in Figure 6B are not exactly the same as would be predicted by an analyses with a "dry" free-draining base, but are very similar. The neutral excess pore water pressure axis is located in a significantly different location from the neutral total pore water pressure axis. The predicted volume of discharge to the impoundment surface and foundation under these two different base boundary conditions would be quite different. These results indicate the significance of boundary conditions to the analyses.

Using the total pore water pressure isochrones of Figure 6A as well as the discharge curves of Figure 5 and the computer output data giving elemental volume changes throughout the profile, the discharge curves representing volume flow rates to the surface and impoundment foundation of Figure 7 were determined. The results given in Figure 7 reflect an important factor; for boundary condition 2, over 80 percent of the total consolidation seepage occurs to the impoundment surface.

Theoretical Considerations - Gravity Drainage

Gravity drainage of interstitial tailings water will occur from the onset of deposition in the impoundment if a "dry" boundary condition exists at the base of the impoundment. Because waste impoundments are typically designed to have very low seepage rates, it is more typical for deposition to occur under saturated conditions with a hydrostatic condition existing at the base of the impoundment during primary consolidation. Under these conditions, gravity

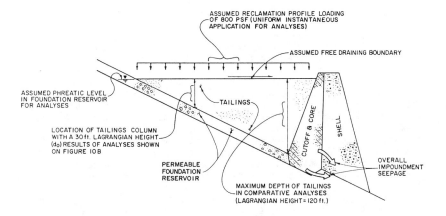

FIGURE 6A: TAILINGS IMPOUNDMENT
CROSS SECTION

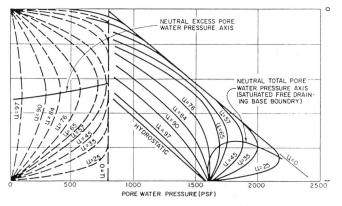

d/d_{tu} = TOTAL PROFILE DEPTH AT INDICATED DEGREE
OF CONSOLIDATION (U %).

— — — EXCESS PORE WATER PRESSURE ISOCHRONES

———— TOTAL PORE WATER PRESSURE ISOCHRONES

FIGURE ILLUSTRATED IS FOR A TAILINGS PROFILE
SUBJECTED TO AN INCREMENTAL LOADING OF 800 PSF.

FIGURE 6B: TOTAL AND EXCESS PORE WATER PRESSURE ISOCHRONES

drainage will not occur until consolidation excess pore water pressures have dissipated. Once excess pore water pressures have dissipated, gravity drainage will occur when the water level in the foundation reservoir falls below the level in the tailings or as a result of the overall seepage from the impoundment.

Gravity drainage that occurs as a result of a "dry" boundary condition is generally associated with tailings desaturation. This is the typical condition for an arid environment tailings impoundment where gravity drainage generally occurs under partially-saturated flow conditions. Gravity drainage as a result of a slow decrease in the water level in the impoundment foundation or as a result of low foundation seepage rates is generally associated with impoundments constructed in relatively wet climates. Gravity drainage in these environments generally occurs under saturated flow conditions.

To analyze the rate of gravity drainage requires, first, an analysis of the overall seepage rate from the impoundment as a function of the driving head behind the embankment. When evaluating an existing impoundment, actual seepage rate measurements over a wide range of driving head conditions may be available. A typical set of field measurements of toe seepage associated with the case history is given on Figure 8. An analysis during the design phase could utilize flow nets or analytical methods to make these estimations.

The next step in the evaluation of gravity discharge is to estimate the maximum volume rate which could occur to the impoundment foundation, i.e. no constraint of gravity discharges (flow gradient of one neglecting losses). These estimates can be made with the help of results from the finite strain consolidation analyses. Typical results for our comparative analyses study case 30-foot and 120-foot profiles are given in Figures 9 and 10, respectively. They included the predicted void ratio profiles after completion of primary consolidation under an 800-psf surcharge load and the predicted distribution of permeability throughout the profile.

The lowest permeability of tailings occurs at the impoundment base where the highest density materials are found. The permeability at the base will control the rate of water discharging under gravity drainage conditions. The results given in Figures 9 and 10 are schematically shown for the case history impoundment (see Figure 2 of Part I) in Figure 11. The results and average gravity discharge rate for the discretized impoundment indicate a range of 0.18 (ft/day) for that portion of the impoundment with an average depth of 15 ft, to 0.06 (ft/day) for portions of the impoundment with an average depth of 100 feet. These rates correspond to 40 and 14 GPM per acre of impoundment base area. The data of Figure 8 indicate that the maximum total seepage rate from the impoundment may be as much as 50 GPM. It can, therefore, be concluded that the gravity discharge rate (as high as 40 GPM/acre of base area) will be directly controlled by the overall impoundment total seepage rate (when the total im-poundment base area exceeds 1.25 acres), and the water balance eval-uations will be crucial to determine the level of the water in the foundation reservoir behind the dam. For a typical impoundment with a

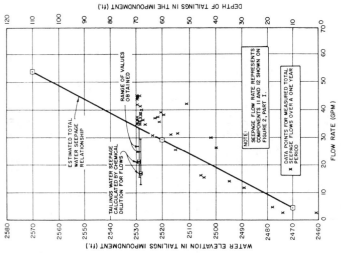

FIGURE 8: ESTIMATED FOUNDATION AND FOUNDATION
CUTOFF SEEPAGE WATER RELATIONSHIP
BASED UPON ACTUAL FIELD MEASURMENTS

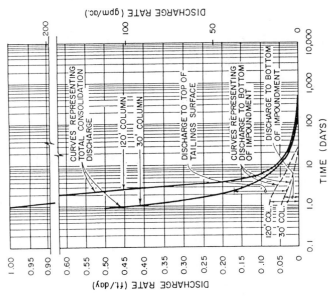

FIGURE 7: CONSOLIDATION DISCHARGE TO IMPOUNDMENT
SURFACE AND FOUNDATION

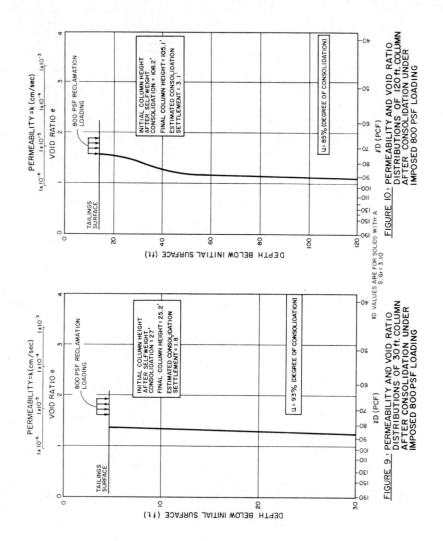

FIGURE 10: PERMEABILITY AND VOID RATIO DISTRIBUTIONS OF 120 ft. COLUMN AFTER CONSOLIDATION UNDER IMPOSED 800 PSF LOADING

FIGURE 9: PERMEABILITY AND VOID RATIO DISTRIBUTIONS OF 30 ft. COLUMN AFTER CONSOLIDATION UNDER IMPOSED 800 PSF LOADING

base area of 200 acres, gravity drainage will be controlled by the tailings permeability when the permeability is on the order of 10^{-7} to 10^{-8} cm/sec. The 10^{-8} value would be appropriate when the overall seepage rate is on the order of one to ten GPM.

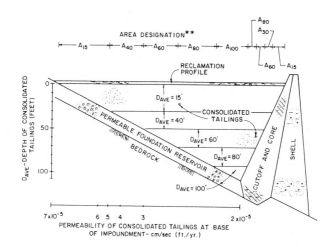

FIGURE II : GRAVITY DRAINAGE RATES

AREA DESIGNATION**	AVE. PERMEABILITY			AVE. GRAVITY DISCHARGE*	
	cm/sec	ft./yr.	ft./day	rate gpm/ac	ac-ft/ac-day
A_{15}	6.2×10^{-5}	64	.18	40	.18
A_{40}	3.8×10^{-5}	39	.11	24	.11
A_{60}	2.9×10^{-5}	30	.08	19	.08
A_{80}	2.4×10^{-5}	25	.07	16	.07
A_{100}	2.1×10^{-5}	22	.06	14	.06

* ASSUMING A FLOW GRADIENT OF 1
** SUBSCRIPT INDICATES AVERAGE TAILINGS DEPTH IN DESIGNATED AREA

In order to complete the water balance, an iterative evaluation of inflows to the foundation reservoir is necessary. One product of the iterative balance will be the actual quantity of gravity drainage. The importance of modeling seasonal variation in inflows to the foundation reservoir other than that from gravity drainage can clearly be seen from the preceding discussion.

Comparing the results of the primary consolidation discharge rates to those from gravity drainage indicate that each aspect of the overall seepage problem is important over different portions of the overall time frame where contaminated seepage may occur. The volume of seepage from gravity drainage that occurs under "dry" boundary conditions (tailings desaturation) has another important aspect.

Specific retention (sometimes referred to as field capacity) is the water which is retained in the material against the forces of gravity. The problem of evaluating specific retention effects on the volume of gravity seepage may have the additional complication of widely varying tailings materials both laterally and with depth in the impoundment. Using the results of the finite strain analyses, total available volumes of water under gravity drainage conditions (tailings desaturation), were estimated. The discretized profile of Figure 11 was evaluated for several different estimated specific retention levels and a summary of results is given in Table 1.

TABLE 1
SUMMARY OF AVAILABLE GRAVITY DRAINAGE WATER VOLUMES

Designated Area	Depth of Layer (ft)	Total Available Water Volume (ac-ft/ac)*			
		SR**=15%	SR=20%	SR=25%	SR=30%
A_{15}	30	11.7	9.9	8.1	6.0
A_{40}	20	6.6	5.2	4.0	2.6
A_{60}	20	6.6	5.2	3.8	2.2
A_{80}	20	6.2	4.6	3.2	1.8
A_{100}	20	6.2	4.6	3.2	1.8

*   The volume of water which drains under force of gravity.
**   Specific retention is the percentage (by weight of dry solids) of water retained against the face of gravity.

The effect of specific retention on predicted volumes of interstitial water discharge becomes more significant with depth. The difference between 15-and 30-percent specific retention within the A_{15} profile is a factor of two where the difference is a factor of 3.4 for the A_{100} profile. The selection of the specific retention used in the analyses can have a significant effect on the results of water balance evaluations. The lack of specific information on many of the different tailings would suggest that additional research is required in order to refine seepage evaluations.

A significant aspect of gravity drainage has been neglected in the preceding discussion. Gravity drainage (for tailings desaturation conditions) will result in increased effective stresses through the tailings impoundment. The net effect will be additional primary consolidation and water discharge to the base of the impoundment, as well as upwards into the partially saturated tailings. The analysis of this phenomenon is complex and current available finite strain computer programs are not capable of analyzing this problem. The water balance approach to modeling water levels in the impoundment foundation reservoir will facilitate the development of computer programs to solve this complex problem in the future.

Theoretical Considerations - Secondary Consolidation

Little is known of the intergranular viscous phenomenon (known in soil mechanics as secondary consolidation) which occurs in tailings having a low density, in an environment of relatively low effective stress. The available theories for predicting vertical settlement, changes in material properties, and possible discharge of interstitial tailings water as a result of secondary consolidation are approximations only (Zeevaert, 1983). The intergranular viscous phenomenon illustrated in Figure 12 is the basis of the available numerical theories. The data in Figure 12 is generally gathered from one-dimensional laboratory oedometer tests.

The slope of the secondary consolidation curve is known as the secondary consolidation index, C_t. On a logarithmic scale, C_t is equal to the measured vertical displacement.

Standard oedometer tests completed on the tailings material being used in the preceding comparative analyses indicated a C_t value of 1.31×10^{-4} (ft) is valid over a range of vertical effective stresses up to 8,500 psf. The results presented in Figure 7 indicate that primary consolidation in the 30-foot profile is complete approximately 100 days after application of the 800-psf surcharge loading. Using the C_t value above, the estimated secondary consolidation settlement over the 100 to 1000-day log cycle would be 0.004 feet. If the assumption is made that settlement during secondary consolidation is directly related to a change in the void ratio, the settlement of 0.004 ft corresponds to an average discharge rate of 4.4×10^{-6} ft/day (1×10^{-3} GPM/acre) over this 900-day period. The results for the 120-foot profile over the 500 to 5000-day log cycle would be 0.016 feet. This equals an average discharge rate of 3.6×10^{-6} ft/day (1×10^{-3} GPM/acre) over a period of 4500 days (12 years).

These results indicate that the most significant rates of discharge occur during the primary consolidation and gravity drainage period. Discharge from secondary consolidation under the most conservative assumptions is several orders of magnitude less than discharge from primary consolidation and gravity drainage, even when gravity drainage is controlled by the overall seepage rate from the impoundment. The results suggest that more sophisticated methods of evaluating secondary consolidation as related to the prediction of seepage from a mine tailings impoundment are not warranted.

Theoretical Considerations - Infiltration

Long-term seepage rates from a tailings impoundment may be affected by infiltration of precipitation and surface runoff water. If water is allowed to impound on the surface of the impoundment after closure, seepage rates resulting from the surface infiltration will be significantly larger than if no ponding is allowed to occur.

The tailings impoundments being designed and constructed in recent years have, in most likelihood, conceptual reclamation programs included in permit stipulations and bonding requirements. Although a limited amount of reclamation activities have been undertaken on tailings impoundments in recent years in "dry" climate areas, extensive construction experience and monitoring data is lacking. The actual implementation of reclamation programs will allow for refinement

of planning and designs. Until then, evaluations of seepage as a result of surface infiltration will be based upon reclamation concepts.

A typical concept of reclamation for a tailings impoundment is
illustrated in Figure 13. The reclamation profile includes a layer of
topsoil, subsoil, and low-permeability cap overlying a cover drain
layer. The thickness of the topsoil and subsoil layers are generally
based upon the estimated depth of root penetration of plant species
to be used. The low-permeability cap is designed to impede soil cap
percolation and allow greater soil moisture availability for plant
growth. It is placed below anticipated root penetration depths in
order to maintain low-permeability characteristics over the long
term. The cover drain layer generally will serve two functions.
First, it will provide a capillary barrier and prevent the migration
of contaminants up into the soil profile where they can be absorbed
by plants and passed on to grazing animals. Second, it will provide a
free-draining boundary at the tailings surface where consolidation
discharge can occur without retarding the consolidation process. The
cover drain layer can be sloped to drain to a positive gravity
discharge point. During the primary consolidation phase, waters
draining from the cover system can be collected and treated or
disposed of in an acceptable manner.

The calculation of an overall infiltration rate is based upon a
water balance principle. The general components of the profile water
balance include precipitation, runoff, infiltration, evapotranspiration, cover drain discharge, and tailings infiltration (recharge).
These components are illustrated in Figure 13.

The estimation of the runoff and infiltration components require
consideration of soil type, slope, runoff impedance (through rill
construction or vegetation establishment), and seasonal climatic conditions. There are two basic methodologies which can be used to
estimate the components of runoff and infiltration:

    1)   Partial saturation flow models, and
    2)   Simplified consumptive use models.

Partial saturation flow models (Walski, et al, 1983 and Schroeder, et
al, 1983) focus primarily upon the flow characteristics of the reclamation profile soils. These models are generally computerized and
include consideration of evapotranspiration, and the specific retention of the soils. They depend heavily upon the results of complex
laboratory tests. The simplified consumptive use models have been developed from theoretical predictions that have been modified to correspond to actual field measurements in a variety of climates. The
primary applications have been to determine "consumptive use" of
water for agricultural irrigation, and to establish water rights.
These simplified methods are based upon soil and plant species
categories and do not require extensive field or laboratory analyses.

One of the most complex processes which occurs in a reclamation
profile after establishment of vegetation is that of evapotranspiration. This factor, as well as the surface runoff component, are gen-

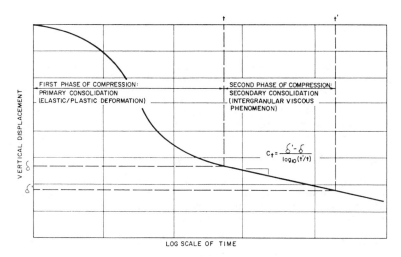

$$C_t = \frac{\delta' - \delta}{\log_{10}(t'/t)}$$

FIGURE 12: GENERAL OBSERVED CONSOLIDATION
BEHAVIOR OF FINE GRAINED TAILINGS

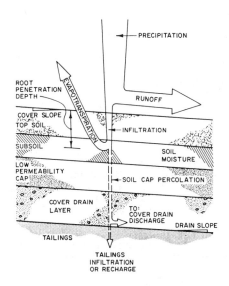

FIGURE 13: TAILINGS IMPOUNDMENT COVER
SYSTEM WATER BALANCE COMPONENTS.

erally the most significant considerations in infiltration evaluations. A comprehensive review and comparison of available methods to
calculate evapotranspiration has been presented by Jensen (1973).
The methods are classified as humidity, radiation, temperature,
combination and miscellaneous. As recently as 1975, the EPA endorsed
the Thornthwaite temperature method (Fenn, et al, 1975). Results of
the study by Jensen (1973) showed that this method generally under
predicts evapotranspiration, often up to 50 percent. Although the
significance of this finding has resulted in other methods receiving
greater endorsement for consumptive use calculations, the
Thornthwaite method would be generally conservative from the standpoint of the amount of infiltration and recharge which may occur and
impact seepage quantities from a tailings impoundment.

A simplified evaluation illustrates the significance of the infiltration component. The Thornthwaite method (1955, 1957), adopted
for the prediction of leachate generation from solid waste disposal
sites (Fenn, et. al., 1975), was used for the evaluation.

The method involves conducting a monthly water balance analysis
for the active soil storage zone which extends to the full depth of
root penetration (shown on Figure 13). Monthly input to the soil
moisture storage is calculated from the monthly precipitation minus
an assumed quantity of runoff (25%). Evapotranspiration losses are
accounted for and are a function of the potential evapotranspiration
and the amount of soil moisture in storage. Whenever the soil moisture storage zone is full to field capacity, downward percolation
occurs. The results of the analysis are in terms of monthly percolation rates which can be adjusted for input to the overall impoundment
water balance evaluation. A summary of the average annual water
balance evaluation for the reclamation profile is as follows.

Inflow -	Precipitation		65.1 inches
Outflow -	Runoff	16.3 inches	
-	Actual evapotranspiration	19.3 inches	
-	Percolation	29.5 inches	
		$\sum$ 65.1 inches	$=\sum$ 65.1 inches

Estimations of actual infiltration to the tailings were made based
upon the relative permeability of the cover drain, slope and the
average length of flow path from the point of percolation to the outlet of the cover drain system. The estimated average rate of infiltration ranged from 10 to 30 percent. This corresponds to three to
nine inches over a period of roughly five to six months.

These values of infiltration correspond to volume rates of
approximately 0.3 to 0.9 GPM/acre. This rate of infiltration is actually quite high and reflects the influence of climatic conditions on
seepage from an impoundment. An impoundment with a net surface area
of 100 acres would have a potential for 30 to 90 GPM of infiltration. In the comparative analyses described in the preceding sections, the overall maximum seepage rate from the impoundment was 55
to 60 GPM. These results indicate that infiltration rates would be
potentially high enough to keep the tailings in a saturated condition

for a very long period of time in this wet environment. They also indicate that predicting gravity drainage and overall infiltration rates requires careful evaluation of climatic conditions.

Summary and Conclusions - Part II

Theoretical considerations for predicting seepage from a mine tailings impoundment due to consolidation (primary and secondary), gravity drainage, and surface infiltration have been discussed. Table 2 is a summary of comparative and illustrative examples presented.

TABLE 2
SUMMARY OF PREDICTED SEEPAGE RATES

Activity	Predicted Range of Interstitial Tailings Water Seepage or Discharge Rates (GPM/acre)
1. Primary Consolidation	*200 to .45
2. Gravity Drainage**	0 to 40
3. Secondary Consolidation	.001
4. Infiltration***	.3 to .9

* Maximum rate is for a 120-ft profile at day 1 following application of 800-psf surcharge load. Minimum rate is at day 300.
** Rates must be compared to the overall seepage rate from the impoundment. The lowest rate will control the rate of gravity drainage.
*** Rates are applicable for a total impoundment area up to approximately 90 acres. For a larger impoundment actual rates would be controlled by the overall maximum impoundment seepage rate.

The following conclusions can be made from the summary results presented in Table 2 as well as the discussions in the preceding sections.

1. The most significant rate of discharge of interstitial tailings water occurs during the initial stages of primary consolidation. The relative time frame of primary consolidation discharge is short compared to the other discharge mechanisms considered.

2. The selection of a consolidation theory for predicting discharge volumes and rates is of primary importance to the evaluation of seepage from a tailings impoundment. We believe finite stain theory to be the best available technology for evaluating the consolidation aspects of mine tailings impoundments.

3. Boundary conditions assumed for consolidation analyses have a significant effect upon the rate of discharge as well as the prediction of the volumes which drain to the surface or foundation of the impoundment. For most tailings impoundments, a sat-

urated, free-draining foundation is appropriate. Under these conditions, up to 80 percent of consolidation discharge may occur to the impoundment surface.

4. The consideration of self-weight in consolidation analyses of mine tailings is vital in order to determine discharge rates and volumes. The predicted volume rate of discharge may be as much as ten times greater using a finite strain analysis over conventional theories. The design of drainage measures as part of reclamation programs on the surface of an impoundment should include anticipated drainage rates predicted from finite strain analysis in order not to inhibit the consolidation process.

5. Gravity drainage of interstitial tailings water is one of the most significant and complicated aspects of long-term seepage estimations. Gravity drainage (unlike primary consolidation discharge which occurs under a gradient caused by excess pore water pressures) in most tailings impoundments will be controlled by the overall impoundment seepage rate. A water balance evaluation is essential to determine foundation water levels and gravity drainage rates.

6. Discharge of interstitial tailings water as a result of secondary consolidation is several orders of magnitude less than discharge from primary consolidation and gravity drainage, even when gravity drainage is controlled by the overall seepage rate from the impoundment. These results suggest that more sophisticated methods of evaluating secondary consolidation as related to the prediction of interstitial tailings water seepage from an impoundment are in the majority of cases not warranted.

7. Infiltration of precipitation and surface runoff water into tailings and subsequent seepage is a significant aspect of the overall seepage evaluation problem, particularly in wet climates. Seepage as a result of infiltration can be of a similar magnitude as gravity drainage in any climate that is similar to that assumed in the example problem.

8. Evaluation of infiltration through reclamation profiles requires consideration of the complex evapotranspiration process. Numerous theoretical methods are available to predict evapotranspiration. The selection of a method depends upon the significance of the infiltration component. Simplified procedures such as the Thornthwaite method, which produces conservative results with respect to infiltration (i.e. they under-estimate evapotranspiration and over-estimate infiltration) may be suitable for evaluation of mine tailings impoundments.

The evaluations described in this paper have helped to identify several key areas where additional research and development of modeling techniques could enhance the overall water balance approach to modeling seepage of interstitial tailings water from mine tailings impoundments. The key areas include:

1. Development of methods of evaluating core seepage considering non-uniform rates of rise, available water volumes, material properties and changing pore water pressures of tailings material on the upstream face of an impounding embankment.

2. Development of methods of evaluating primary consolidation of the tailings using finite strain theory which can account for variable foundation boundary conditions and geometrical effects. The methods should include considerations of changing effective stresses as a result of gravity drainage and lowering of the saturation level within the tailings as well as two and three dimensional consolidation effects imposed by the geometry of an impoundment. The development of such methods should be based upon a water balance approach.

3. Additional research on the fundamental characteristics of tailings materials as related to gravity drainage and infiltration. Such research would include specific retention characteristics as well as partially saturated permeability for other than uranium tailings (Martin, et. al., 1980; Veyera, et. al., 1981) which may affect long-term infiltration rates.

4. Research of the actual consolidation behavior within tailings impoundments including measurements of in-situ pore pressures and comparison with predicted values, and measurement of in-situ soil properties such as density and permeability.

References

Biot, M.A., (1941), "General Theory of Three-Dimensional Consolidation", Journal of Applied Physics, 12, pp. 155-164.

Biot, M.A., (1956), "General Solutions of the Equations of Elasticity and Consolidation for a Porous Material", Journal of Applied Mechanics, 23, pp. 91-96.

Caldwell, J.A., Ferguson, K.A., Schiffman, R.L., and van Zyl, D., (1984), "Application of Finite Strain Consolidation Theory for Engineering Design and Environmental Planning of Mine Tailings Impoundments", Proceedings of the ASCE Speciality Conference on Sedimentation/Consolidation Models, pp. 581-606.

Croce, P., Pane, V., Zridarcic, D., Ko, H. Y., Olsen, H. W., and Schiffman, R.L., (1984), "Evaluation of Consolidation Theories by Centrifuge Modeling:", Proc. of the ISSMFE Spec. Conf. on the App. of Centrifuge Modeling to Geot. Design, Eng. Dep., Univ. of Manchester, Manchester, England.

Fenn, D.G., Hanley, K. J., and DeGeare, T.V., (1975), Use of Water Balance Method for Predicting Leachate Generation From Solid Waste Disposal Sites, EPA rept. SW-168, Office of Solid Waste Mgmt. Prog.

Gibson, R.E., and England, G.L., and Hussey, M.J.L., (1967), "The Theory of One-Dimensional Consolidation of Saturated Clays; 1. Finite Non-Linear Consolidation of Thin Homogeneous Layers", Geotechnique.

Jensen, M.E., (1973), "Consumptive Use of Water and Irrigation Water Requirements", A report prepared by the Tech. Committee on Irrigation Water Req. Irrigation and Drainage Div., ASCE, pp. 113-154.

Martin, J.P., Veyera, G.E., Nasiatka, D.M. and Oxorno, L.F. (1980), Characterization of the Inactive Tailings Sites, Proceed. of Symp. on Uranium Mill Tailings Mgmt. Fort Collins, CO, pp. 557-578.

Mikasa, M., (1963), The Consolidation of Soft Clay - A New Consolidation Theory and Its Application, Kajima Institution Publishing Co., Ltd. (In Japanese).

Mikasa, M., and Ohnishi, H., (1981), "Soil Improvement by Dewatering in Osaka South Port", Proceed. of the Sixth Internat. Conference on Soil Mechanics and Foundation Engineering, pp. 639-664.

Schiffman, R.L., Chen, A.T-F., and Jordan, J.C., (1969), "An Analysis of Consolidation Theories", Journal of the Soil Mechanics and Foundation Div., ASCE, 95, No. SM1, Pro. Paper 6370, pp. 285-312.

Schiffman, R.L., Pane, V., and Gibson, R.E., (1984), "The Theory of One-Dimensional Consolidation of Saturated Clays, IV. An Overview of Non-linear Finite Strain Sedimentation and Consolidation", ASCE Specialty Conference on Sedimentation/Consolidation Models: Theories and Predictions, San Francisco, California, pp. 1-29.

Schroeder, P.R., Gibson, A.C., and Smolen, M.D., (1983), "Documentation for the Hydrologic Evaluation of Landfill Performance (HELP) Model," EPA, Cincinnati, OH, 206 pp.

Terzaghi, K., (1923), "Die Berechnung der Durchlassigkeitsziffer des Tones aus dem Verlauf der Hydrodynamischen Spannings scheinungen," Akademic der Wissenchaften in Wien, Sitzungsberichte, Mathematisch-naturwissenschaftliche Klass, Part II a, 132, No. 3/4, pp. 125-138.

Terzaghi, K., and Frohlich, O.K., (1936), Theore der Setzung von Tonschichten; eine Einfuhrung in die Analytische Tonmechanik, F. Deuticke, Leipzig, Germany.

Thornthwaite, C.W. and Mather, J.R., (1955), "The Water Balance, Centerton, N.J.," Drexel Inst. of Technology, Laboratory of Technology, Publication in Climatology, 8, No. 1, 104 pp.

Thornthwaite, C.W., and Mather, J.R., (1957), "Instructions and Tables for Computing Potential Evapotranspiration and the Water Balance, Centerton, N.J.", Drexel Inst. of Technology, Laboratory of Technology, Publications in Climatology, V. 10, No. 3, pp. 185-311.

Veyera, G.E. and Nelson, J.D. (1981), Unsaturated Hydraulic Parameters of Grand Junction Uranium Tailings, Proc. of Symp. on Uranium Mill Tailings Management, Fort Collins, Colorado, pp. 557-578.

Walski, T.M., Morgan, J.M., Gibson, A.C., and Schroeder, P.R., (1983), "User Guide for the Hydrologic Evaluation of Landfill Performance (HELP) Model," EPA, Cincinnati, Ohio, 152 pp.

Zeevaert, L., (1983) Foundation Engineering for Difficult Subsoil Conditions, Van Nostrand Reinhold Company, pp. 85-104.

CONTAINMENT OF TEXTILE WASTE
USING A GEOMEMBRANE

By Michael W. Bowler[1]

ABSTRACT

A wastewater treatment plant for a textile mill incorporated two 2.7 million gallon aeration ponds which had originally been lined with clay. The clay liner leaked and caused the formation of sinkholes in the limestone substrata; this necessitated the redesign of the liner with a specified leak rate of 2.7 gallons/min/pond to prevent further sinkhole formation. The new design called for a synthetic membrane liner with a full leak detection system so that any holes could be detected and, if necessary, repaired immediately. The liner chosen was High Density Polyethylene with an extrudate fusion welding system for the seaming. Each pond had four concrete structures in the form of two outlet pipes, an inlet and an access slab. The fixing of the liner to these structures was accomplished with stainless steel battens, anchor bolts and expanded neoprene gasketing. The installation was carried out with a strict quality control program for the seaming, membrane integrity and bolting detail to structures. The result after filling the ponds during the tests period was a leak rate that was too small to measure.

INTRODUCTION

A small city in Alabama operates a waste water treatment plant specifically for the treatment of effluent from a textile manufacturer which consists of waste water from a yarn mill's knitting, dyeing and finishing operations (cotton and polyester), with pH and temperatures in the range of 12.0 and 120° F respectively.

Incorporated in the treatment plant are two 2.7 million gallon aeration ponds using three floating aerators in each pond. The aerators are moored in position using cables and anchor poles embedded at the top of the side slopes. Each pond is 13 feet deep and measures 167 ft. by 357 ft. at the top of sideslope; the sideslopes are at three horizontal to one vertical. Each pond has three penetrations in the form of pipes with concrete collars: an inlet, an outlet surrounded by an aluminum baffle (to hold back

---

[1]Synthetic Linings Manager
Geo-Con, Inc., P.O. Box 17380, Pittsburgh, PA 15235

floating solids) and an emergency outlet in case the pond needs to
be emptied quickly.

During the original design stage, investigation of the local
geological conditions indicated that the proposed pond location was
over a limestone formation with a potential for the formation of
sinkholes. The toxic nature of the effluent and the potential for
sinkhole formation dictated the need for a liner for the ponds. The
liner chosen was 24 inch natural clay on the bottom and sideslopes
of the ponds, in addition the top six inches of the bottom clay was
mixed with bentonite to ensure a permeability of 10^{-7} cm/sec.

Despite the care taken with the installation of the clay liner,
large sinkholes formed in the base of the ponds within days of
filling the ponds. This necessitated the emptying of the ponds and
a reevaluation of the liner system.

LINER SYSTEM REDESIGN AND PERFORMANCE CRITERIA

The primary requirement of any new liner system was that no effluent
could be allowed to permeate the subgrade and cause further sinkhole
formation. With this in mind, it was felt that the clay liner by
itself, even if it was reconstructed, could not offer sufficient
reliability. The new design was in three parts, the first being to
fill the sinkholes and repair the clay liner, the second to provide
a leak detection/collection system and thirdly to install a
synthetic membrane as the primary liner. The leak detec-
tion/collection system had two functions. First to alert the
operator of the plant that there was a leak so that it could be
repaired. Secondly to offer a path of least resistance so that any
leakage through the synthetic membrane could flow to a collection
point without penetrating the clay liner.

The main performance criterion for the synthetic membrane was that
it could leak no more than 1 gallon/million gallons, this meant that
each pond was allowed to leak no more than 2.7 gallons/min. This
leakage rate was determined as being acceptable as it could easily
be drained by the leak detection/collection system. It was reasoned
that some leakage had to be allowed as it would be virtually
impossible to attain zero leakage for ponds of this size.

(a) Clay liner reinstatement

The specification called for filling the inaccessible voids between
large boulders at the bottom of the sinkholes with concrete and then
backfilling the sinkholes with 6 inch compacted layers of local
material to within two feet of top of subgrade level. The clay
liner was then to be rebuilt using the same type of clay and
bentonite as was used in the original clay liner construction.

On completion of the clay liner repair, all vegetation in the
impoundment areas was to be stripped and the surface rolled to
present a smooth hard surface suitable for the laying of a synthetic

membrane. Finally the area was to be sprayed with a herbicide to
prevent vegetative growth under the liner. To facilitate access
into the ponds one concrete access slab per pond was to be built in
the sidslope.

(b) Leak detection/collection system

On completion of the clay liner, the bottom of the ponds were to be
filled on the outer edges with clay, to form a grade of 2% towards
the center of the pond. The grade would allow effluent to drain to
a low point to be collected and would also allow any gas generated
under the liner to flow to the sideslopes without forming pockets
under the liner. Along the center line and through the end dike of
each pond a subsurface drain was to be constructed terminating at
the low end in a collection manhole and at the upper end as a gas
vent through the liner. The subsurface drain along the bottom of
each pond was to be a trench filled with crushed stone encasing a
perforated pipe, the trench grade to be 0.2% towards the manhole.
The pipe to be unperforated where it cut the dike and up the
opposite end sideslope. This pipe system was to be capable of
collecting effluent and draining it to a point where it could be
collected and treated, and also collecting any gas buildup and
venting it through the liner at a point above the effluent level.
For details of the leak detection collection system see Fig. 1.

Finally, over the already prepared surface, a geotextile was to be
laid, the purpose of the geotextile being:

- to enable any leakage through the membrane liner to drain in the
  plane of the fabric to the leak collection drain.

- to enable any generated gas to flow upwards to the gas vents at
  the top of the sideslopes.

- to act as a protective layer between membrane liner and
  subgrade.

- to offer support to the liner in the event of localized subgrade
  movement.

A further requirement of the fabric was that it be capable of
withstanding any chemical attack from the contained effluent. The
geotextile selected that best suited this application was a 110 mil
thick polypropylene needle punched, nonwoven fabric with high
transmissivity in the plane of the fabric.

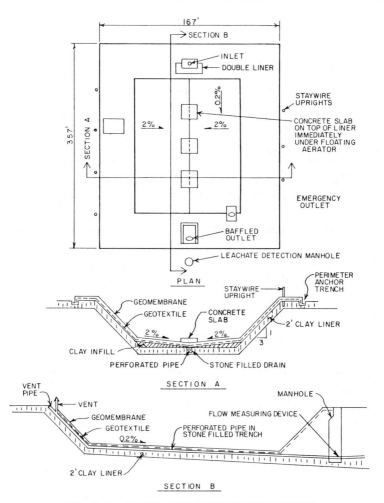

FIGURE I GENERAL LAYOUT AND SECTIONS OF POND

(c) Membrane liner

The geomembrane was to be laid directly on the geotextile.

Requirements for the liner were as follows:

- resistance to the contents of the pond.

- resistance to the high effluent temperature, and also to ambient temperature fluctuations.

- resistance to ultra-violet degradation as the liner was to be left exposed. An exposed liner will facilitate any repairs if they are ever necessary.

- that it be capable of handling local tensile stresses due to differential movements in the subgrade.

- resistance to damage from wave action, surface runoff and personnel working an aerators, etc.

- it must have seams that are capable of handling all the conditions that the sheet itself is capable of handling.

- that seams including factory seams to be tested to prove their watertightness.

- that no more than 2.7 gallons/min/pond leakage could be achieved as measured in the leak detection pipe.

- that it be capable of being fixed to concrete structures with no leakage.

The liner selected was a non reinforced 36 mil High Density Polyethylene which met the design requirements for the membrane.

LINER INSTALLATION

Prior to installation a sheet layout was drawn up, indicating where all seams were to be placed.

The purpose of the layout plan was to optimize sheet sizes, to minimize weld length, to make sure that welds were not in tension (particularly on the sideslope where all welds must go up and down the slope and not across). In the field, the layout plan was used as an as-built drawing by indicating the location of patches and relating sheets to the manufacturer's roll numbers.

The first phase of construction was to excavate the anchor trench at the top of the sideslope placing the excavated material away from the pond. Next came the laying out of geotextile and geomembrane, both operations used the same equipment and crew, and were carried out in conjunction with each other. The placing of fabric and membrane was at a rate no faster than that of the seaming crew. (See Fig. 2.)

FIG. 2 LAYING OF GEOTEXTILE AND GEOMEMBRANE

All sheets of fabric and membrane were buried in the anchor trench
(see Fig. 3). All loose, unseamed edges were sand bagged to prevent
wind uplift. The geotextile was not seamed or stitched in any way
but was overlapped a minimum of twelve inches to ensure that there
would be no restriction of flow in the plane of the fabric.

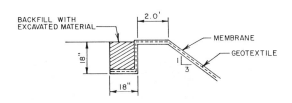

FIGURE 3   ANCHOR DETAIL

The membrane was overlapped four inches and taped together using a double sided tape. This was to hold the two sheets together during the seaming process. The area to be welded was then cleaned thoroughly. The area where the weld was to be placed was treated using a grinder to remove the top film of oxidized material.

The seaming process used was a fusion/extrusion weld. The seam consists of a hot extrudate of the same material as the liner being placed, under pressure, over the overlap of the two sheets. This causes the two sheets to fuse together to form a true homogeneous weld. (See Fig. 4.)

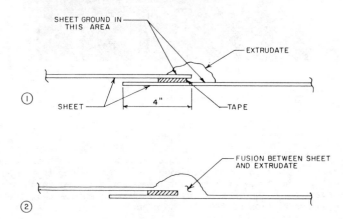

FIGURE 4   WELDING DETAIL

All welding was according to the sheet membrane manufacturer's welding manual which laid out parameters for welding machine settings, extrudate temperature settings, weld area preparation and acceptable weather conditions. The seaming was carried out by welding technicians who were trained by the membrane manufacturer and had a minimum of 2 million square feet of welding experience.

The liner installation was subject to full quality control and any work that was not to the Engineer's and Manufacturer's specifications was repaired and retested.

Where the staywire supports protruded through the liner at the top of the embankment, boots were fabricated to prevent rain or flood water from getting beneath the liner. (See Fig. 5.)

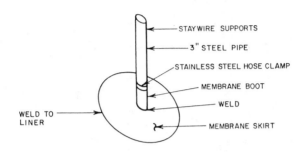

FIGURE 5   STAYWIRE SUPPORT SEAL

The liner was fixed to concrete structures using a stainless steel batten with anchor bolts and neoprene gasketing above and below the liner. (See Fig. 6.)

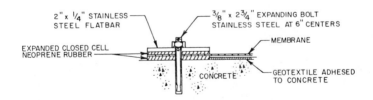

FIGURE 6   BOLTING DETAIL

At the inlet the liner was doubled to provide protection in the event that there are any solids in the liquid as it enters the pond. All the pipe penetrations were surrounded by concrete to which the liner was attached by batten and bolts. (See Fig. 7.)

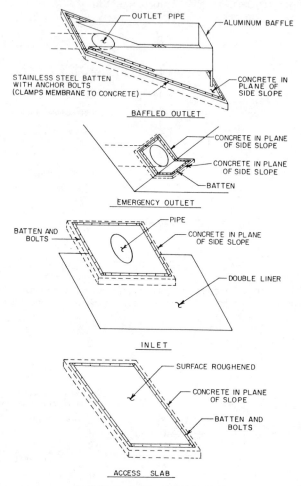

FIGURE 7 PIPE PENETRATIONS

Gas vents (See Fig. 8) were placed at fifty feet intervals around the perimeter of the ponds above the liquid level.

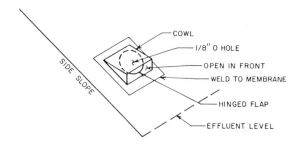

FIGURE 8   VENT DETAIL

When the liner was complete, three concrete slabs were placed on top of the liner directly beneath where the aerators would be positioned. (See Fig. 1.) This was to prevent possible liner uplift due to vortex action of the aerator or in case the aerator had to be lowered and repaired in place.

QUALITY CONTROL FOR LINER INSTALLATION

1. A full time Quality Control technician was on site at all times when seaming was being carried out.

2. Quality Control of the welding was according to the manual supplied by the manufacturer. This manual established such parameters as welding machine settings, temperature settings, prior to weld taping and preparation and acceptable weather conditions.

3. Prior to commencing the day's welding a six foot test piece was welded and tested in shear and peel to determine if the weld was acceptable.

4. Random samples were cut from in-situ welds and tested in shear and peel, the criterion for this test was that the liner fail before the weld.

5. Any welds that visually caused some concern were repaired.

6. All welds were vacuum box tested to ensure that there were no holes. This test utilized a plexiglass topped box placed over

the weld that had previously been wet with a soapy solution.
When air is withdrawn from the box, bubbles will form if there
is a leak as air is sucked through any holes.

7. At the end of the complete liner installation, the liner was
squared off using chalk lines and each square visually checked
for holes, scuff marks, cuts and any other damage. The squares
enabled a systematic checking of the whole liner, with no areas
being overlooked.

8. All drilling and bolting details were checked for tight bolts
and that the gasketing was sufficiently compressed between bolts
to ensure sufficient sealing.

HYDROSTATIC TESTING

The ponds were filled with water and the aerators fixed in position
on completion of all work, quality control and repairs. The leak
rate was monitored for a period of two weeks in the manholes where
measuring weirs had been positioned for this purpose. The acceptable leak rate was 2.7 gallons/min/pond or less. (See Fig. 9.)

FIG. 9 HYDROSTATIC TESTING

RESULTS

(a) Quality Control

   (1) No insitu weld samples taken from both ponds failed in
       shear or peel. This result showed that a true fusion
       between top and bottom membrane sheet had occured.

(2) Vacuum box testing of the total 7,000 feet of weld revealed a total of 19 pinholes.

(3) Visual inspection of the welds found 21 points that caused some concern regarding the alignment of the weld relative to the overlap's position i.e. the weld was not centrally located above the overlap.

(4) Visual inspection of the sheet located 62 holes that had been caused during the installation.

(5) Visual inspection of the bolting detail revealed no potential problems.

(b) Hydrostatic testing:

After two weeks of testing under a full head of water , neither pond leaked enough to produce flow over the weirs placed in the leak detection pipes.

FIG. 10 AERIAL VIEW OF COMPLETED PROJECT

CONCLUSION

The success of this project shows that it is possible to specify and construct geomembrane lined impoundments with very low leakage rates. A liner project of this standard can only be achieved through a design that anticipates all potential problems and an installation with strict quality control to maintain the required standards.

REFERENCES

1) Wallace, R. B. and Eigenbrod, K.D. "An Unprotected HDPE Liner in a Subarctic Environment" Proceedings-International Conference on Geomembranes, Denver, 1984.

2) Kastman, Kenneth H. "Hazardous Waste Landfill Geomembrane Design, Installation and Monitoring" Proceedings-International Conference on Geomembranes, Denver, 1984.

SEEPAGE CONTROL FOR EMBANKMENT DAMS
USBR PRACTICE

by: C. Leo Mantei,* Member ASCE

ABSTRACT: The current Bureau of Reclamation approach to seepage assessment, analysis, and design for embankment dams is described. The advantages and appropriate applications for the analysis programs in use are discussed, as are methods and sources of material property data and criteria for safe design. Seepage control features are reviewed, and brief references to appropriate case histories are made.

INTRODUCTION

It is appropriate to emphasize seepage control in embankment dam design. Studies (14) indicate that piping and seepage were responsible for 38 percent of all embankment dam failures from 1900 to 1975.

This paper describes seepage analysis and control for embankment dams as practiced at the Bureau of Reclamation, Engineering and Research Center in Denver. The subject is discussed under the following headings:

1. Seepage Assessment,
2. Seepage Analysis Methods,
3. Material Properties,
4. Failure Modes, and
5. Seepage Control Features.

Examples in Bureau practice are referred to briefly to illustrate the seepage concerns being discussed.

SEEPAGE ASSESSMENT

For most new projects, the design data gathering period prior to initiation of final design is the time when the potential for seepage-related problems is revealed.

For existing projects, the current Safety of Dams Program provides an orderly and time-controlled process for the expedient assessment and resolution of seepage and other critical concerns.

Most serious seepage problems initiate from deficiencies in the natural foundation that have the potential for destroying the integrity of the dam and reservoir. Therefore, in addition to

---

* Civil Engineer, U.S. Department of the Interior, Bureau of Reclamation, D-230, PO Box 25007, Denver, CO, 80225-0007

detailed logging, sampling, and testing of the foundation, borehole
permeability testing of continuous intervals in a large percentage of
the drill holes at each site has been a Bureau practice for many years.
The usual indicators of potential seepage problems from this type of
investigation are:

1. Jointed, solutioned, or moderately weathered formation rock.
2. Coarse-grained alluvial or fluvial deposits.
3. Water-soluble minerals.
4. Significant water-take during permeability testing.
5. Dispersive, or otherwise undesirable embankment construction
materials.

    At existing dams, some of the additional indicators observed to
detect seepage problems are:

1. Unusual piezometer readings in embankment and/or foundation.
2. Dampness or flow from the embankment, other than from drains.
3. Unusual deformation of the embankment.
4. Suspended soil in any seepage effluent, controlled or
uncontrolled.
5. High or unusual changes in seepage exit flow rates.
6. Lush vegetation.

    When serious seepage problems are suspected, additional, and
sometimes more sophisticated methods are employed. These include, but
are not limited to:

1. Chemical comparison of reservoir water with seepage effluent or
ground-water samples.
2. Dye and radioactive tracer studies.
3. Satellite and aircraft imagery - including photography and
infrared scanning.
4. Geothermal survey - relating temperature anomaly to seepage
velocity.
5. Resistivity survey - relating electrical conductivity to
saturated soil.
6. Self-potential survey - relating induced voltage gradient to
seepage velocity.
7. Special laboratory testing of undisturbed samples.
8. Numerical seepage modeling - described later in this paper.

SEEPAGE ANALYSIS METHODS

Equation and Chart Methods
     The basic Darcy's Law equation, and the extensive combination of
formulas, tables, and charts assembled under the sponsorship of the
Corps of Engineers, U.S. Army (16)(17)(19) are sometimes used for preliminary and final design of seepage control features such as relief
wells, trench drains, blankets and berms when more sophisticated
methods are not justified by complex conditions or extensive data.

Graphical Methods
     The phreatic surface and flow net procedures described by
Casagrande (8) and others (9)(15)(25) are occasionally used to study

the general distribution of pore pressures in an embankment or foundation, when conditions can be simplified sufficiently. However, anisotropy, multiple materials, geometric complexities in design, the
need for parameter variation, and the need for graphical skill and
experience limit the purely graphical procedure for routine use,
except in the simpler cases.

Analogy Methods

The electric analogy method was developed to a highly refined
state and existed as the Bureau's primary seepage analysis tool (51)
for over 30 years. The most popular form was the electrolyte tank
wherein the porous medium was represented by tap water at various
depths to account for differences in permeability. Many two- and
three-dimensional models of embankments, concrete dams, and appurtenant structures were studied for phreatic surface, uplift pressure,
exit gradient, flow quantity, and the effectiveness of underdrains and
seepage cutoffs. The electric analogy is advantageous over flow net
methods for three-dimensional geometries, but has many of the same
weaknesses.

Numerical Methods

Finite Difference Method - The Bureau does not use the finite
difference method directly for solving localized seepage problems, but
does use planar versions of the method for regional ground-water
analysis in planning and operations studies. In one study (12) the
results of a regional finite difference model were used to establish
boundary conditions for a three-dimensional finite element site model
which is described later.

Finite Element Method - The finite element method in two and three
dimensions has been used for seepage analysis at the Bureau since 1977.
When the transition from electric analog to numerical models began,
several dual model sets were used to establish good numerical model
practices in terms of appropriate mesh refinement, accuracy of flow
quantities, and accuracy of flow and pore pressure distribution. The
electric analogy is justified for this type of verification because its
validity as a model for saturated porous media flow is well documented,
and in numerical model terminology it represents a model of infinite
mesh refinement.

Many two-dimensional vertical sections through dam and/or foundation have been studied (22) using saturated flow finite element (45)
simulation in the steady-state mode. Most of these models were developed to study seepage occurrence related to permeable foundations;
first without proposed seepage control features and then, if
necessary, with various combinations of seepage control features which
are described later.

Three-dimensional finite element (44) seepage models have been
used by the Bureau for several projects.

1. Proposed Narrows Dam (5) - Localized model to determine the
magnitudes and extent of uncontrolled exit gradients in the sandy
foundation at the longitudinal termination of a slurry trench
cutoff, and the effectiveness of relief wells as a counter-measure

2. McPhee Dam (1) - Localized model to study phreatic surface and
seepage flow in a sandstone abutment, and a site model to study
reservoir seepage losses

3. Calamus Dam - Staged modeling (12) from regional to site to
localized models for determining reservoir losses, distribution of
seepage into drains and other exit points, dewatering requirements
(50), effectiveness of drainage alternatives, and design data in
the form of seepage gradients and flows

4. Several Quasi-Three Dimensional Models (43) - To study relief
well spacing at the toe of a dam. The side boundaries were assumed
to be "no flow" boundaries corresponding to the line of symmetry
between wells, and the line of symmetry at a well

A recent addition to the Bureau's finite element capability in
seepage analysis has been the variable saturation model (33). It has
the obvious advantage over saturated flow models of being able to
simulate the moisture effects in the whole system rather than just the
saturated portion. Because of this, discontinuous zones of saturation
can be modeled and the finite element mesh remains invariant during
transient analyses, resulting in more efficient and accurate computation of pore pressure, etc., especially during reservoir fluctuation. The program currently being used by the Bureau (21) has the
additional capability of simulating infiltration above the phreatic
surface as well as surface evaporation and moisture consumption by
plants, if necessary. The program has been used (13) successfully by
the Bureau to study:

1. The changes in pore pressure within an embankment due to
transient reservoir and tailwater level changes, and

2. The reduction in moisture content within an embankment due to
gravity drainage and evaporation, related to desiccation cracking.

In addition to the above, the program has additional intended
applications for:

1. Realistic prediction of pore pressures for slope stability
analysis during a rapid reservoir drawdown,

2. Estimating quantitative data for first-fill monitoring, and

3. Studying ground-water mound buildup due to seepage from unlined
canals above the water table under transient canal operation.

The versatility of this program (and method) is not, however,
without its drawbacks and should not be looked upon at this time as a
general purpose tool for all applications. The major items of additional effort required, over that required for a saturated flow model
are:

1. A relatively more refined finite element mesh;

2. Initial condition pore pressure data (both positive and negative) for every node;

3. More exotic data gathering and input preparation for variable material properties (hydraulic conductivity and negative pore pressure vs. moisture content); and

4. If the climatic and plant growth options are exercised, appropriate evaporation, transpiration, and precipitation data must be assembled.

Boundary Element Method - A recent addition to the Bureau's analytic tools for seepage is the Boundary Element Method (11) which offers greater computational efficiency for solving two-dimensional saturated flow problems in the steady state. The model is discretized only along the boundaries of each material zone, except for additional nodes added for output information. The program used by the Bureau automatically iterates the model to the correct solution for unconfined cases where the position of the phreatic surface is unknown. It is used to solve the same type of problems as the two-dimensional saturated flow finite element program discussed previously, but does not, at present, have a transient capability.

MATERIAL PROPERTIES

Hydraulic Conductivity at Saturation

Hydraulic conductivity and permeability are used interchangeably in this paper, and refer to the in situ flow rate under a unit (1.0) hydraulic gradient.

Primary permeability refers to the intact material, while secondary permeability refers to flow through the fractures or other finite openings in the material. The primary permeability is often negligible compared to secondary permeability in typical fractured rock foundations. During in situ testing under secondary permeability conditions, the state of flow (laminar or turbulent) is unknown. The test is normally interpreted as a laminar flow test, and the quantity computed is assumed to be the equivalent laminar flow permeability. Reference (28) is an excellent discussion of this subject, and offers a method to determine when fracture systems can be treated as porous media.

Sources of saturated hydraulic conductivity data are numerous and varied in degree of dependability:

1. Data from the literature. - This source is often used as a starting point for foundation permeabilities when a geologic log or other description of the materials is available but test data are not available. References (18), (27), (31), and (32) have useful hydraulic conductivity data.

2. Data from design and construction records. - This source of data comes mostly from within the Bureau's project files and includes:

a. Foundation field permeability test data
b. Embankment construction record field permeability test data
c. Remolded laboratory sample permeability test data
d. Gradation and other laboratory data from which permeabilities can be estimated using the Moretrench curves (36)
e. Published Bureau data (3)

Data from this source are used directly for reanalysis of existing
dams.

3. Data from laboratory testing. - Reference (2) describes most of
the laboratory procedures used by the Bureau. Laboratory permeability tests are most useful for testing fine-grained embankment
core or blanket materials, because a large range of gradients can
be imposed, and the higher gradients make shorter test times
possible.

4. Data from field permeability tests. - Four types of in situ
field permeability tests are routinely used by the Bureau to
determine foundation permeability.

a. Aquifer Pumping Tests (4) are the most dependable (and the
most expensive) method of determining the effective horizontal
hydraulic conductivity of a pervious stratum, but are limited to
testing below the water table. They eliminate localized effects
on tests results by measuring reponse over a finite portion of
the aquifer rather than over infinitely small portions of the
aquifer as with other methods.

b. Bore Hole Pressure Tests (designation E-18 of the Earth
Manual (2)) are pump-in tests that apply water pressure to pre-
selected or continuous intervals of the bore hole through the
use of packers to isolate the interval to be tested. They are
performed above or below the water table, and provide good definition of the relative horizontal permeabilities of the various
strata. Because they are performed at many intervals along the
depth of the hole, the test time is short (3-10 minutes normally) and numerical values are subject to questionable
accuracy, especially above the water table where saturation is
incomplete. The pressure applied during these tests must be
predetermined and carefully controlled to prevent hydraulic
fracturing of the formation, which would nullify the test.
Reference (40) provides an excellent update of this method as it
applies to testing above the water table.

c. Shallow Well Pump-in Tests (Chapter III, B of the Drainage
Manual (6)) are used to measure horizontal hydraulic conductivity of surficial deposits above the water table. The composite permeability for the portion of the well that is filled
with water is determined after the pump-in flow rate has
stabilized for a constant head.

d. Ring Permeameter Tests (also Chapter III, B. of the Drainage
Manual (6)) are designed to measure the vertical hydraulic conductivity of surficial deposits above the water table. An

impermeable ring is inserted in the soil to be tested to control
the direction and cross-sectional area of flow. Tensiometers
and piezometers are used to confirm approximate saturation.

5. Hydraulic conductivity data from performance data. - Analysis
of existing structures often has the advantage of performance data
from which the initially selected material properties can be
verified. This process involves calibration of a model by trial
and error changing of critical material hydraulic conductivity
until the model accurately reproduces a known set of steady-state
piezometric and/or flow measurements. This does not yield a unique
solution, however, since there are usually several combinations of
material properties that will produce the same results. Calibration greatly increases the confidence level of the model. Care
must then be taken when using the model for parameter variation
studies that the conditions under which it was calibrated are not
grossly violated.

Hydraulic Conductivity vs. Water Content

For studies of variable saturation effects on embankment dams the
variation of hydraulic conductivity with water content is a required
material property. It is usually expressed as relative permeability
(k_r), a fraction of the saturated permeability (k).

The hysteretic nature of the functional relationship between k_r
and water content (r) is recognized, with higher k_r occurring during
drainage than during infiltration. Freeze (24) and others (34) seem
to give guarded support for the use of a single valued function to
express k_r for practical embankment dam problems.

A single valued curve from the literature (33) for the k_r function has been used thus far by the Bureau in model studies using
program UNSAT2 (21). Future work will make use of modified laboratory
procedures for constant head permeability tests when samples are
available.

An empirical relationship between k_r and the fractional saturation (S) proposed by Corey (20) as $K_r = S^e$ can also be used in the
absence of specific data. The empirical exponent (e) varies approximately from 3 to 6, with low values related to uniform pore size
distribution and high compaction and high values related to well
graded soils containing fines and less compaction.

Soil Suction vs. Water Content

Negative pore pressure (soil suction) as a function of water content is also a soil property that must be evaluated for variable
saturation studies. Like the variable saturation hydraulic conductivity function, soil suction is hysteretic, but can be approximated
as a single valued function for practical applications in embankment
dam seepage studies.

Data from Bureau files were used in early variable saturation
studies by the Bureau. Laboratory procedures (designation E-16 of the

Earth Manual (2)) are available when specific data from samples are needed.

Porosity

Porosity is expressed as the fractional volume of voids in a total volume of soil. It is used as the upper limit of volumetric water content in unsaturated flow studies. Porosity (n) can be determined from the dry unit weight of the soil and the apparent specific gravity of the soil particles. Designations E-23 and E-24 of the Earth Manual (2) describe methods for measuring in situ unit weight of shallow soils. Unit weight of deep soils can be determined from undisturbed sampling.

"Effective porosity" or specific yield is the fractional volume of water that can be drained from the total volume of saturated soil, and is used as the storage coefficient for saturated flow studies. Specific yield is determined in the field from aquifer performance tests (4) and in the laboratory from specific retention tests (6).

References (26), (32), (36), and (37) contain representative data on porosity, and reference (26) contains representative data on specific yield.

FAILURE MODES

The seepage analysis of an embankment dam is not complete unless the model results are compared against criteria by which the factor of safety of the design against failure can be estimated. The most common failure modes for which established criteria are available will be discussed first.

1. Excessive uncontrolled exit gradient will cause soil particles at the toe of a dam to become buoyant. The manifestation of this loss of gravitational stability depends on the composition of the soil in which it is occurring.

   a. In a foundation soil with a high percentage of large particle sizes the fine particles may be removed and deposited on the surface as "sand boils" while the structure of the large particles remains stable, resulting in an increase in permeability and seepage flow.

   b. In a granular foundation soil with a narrow distribution of relatively small grain sizes a mass of soil can become fluidized as the reservoir reaches a hydraulic head coinciding with the critical gradient (I_c) of the soil mass. Rapid catastrophic failure can then result.

   c. A common occurrence is where the foundation soil mass is heterogeneous and may have some cohesion in the surficial layer. Piping will usually start at points of discontinuity or flow concentration in the surficial material such as open drill holes, rodent holes, post holes, root holes, and ditches. As

soil particles are removed, the formation of the pipe(s) pro-
gresses upstream generally following the path of a theoretical
flow net stream line, but wandering to follow weaknesses in the
heterogeneous foundation. The progression of a pipe accelerates
as the flow path from entrance to exit shortens, increasing the
hydraulic gradient.

This type of failure can be gradual, depending on the cohesive-
ness of the soil, and especially if the critical gradient is
only reached intermittently, as in the case of a flood control
dam that only reaches critical gradient during flood storage.

The safety factor (SF) related to exit gradient is usually
expressed as the ratio of critical gradient (I_c) to the vertical
component of the exit gradient (I_e), SF = I_c/I_e. The exit gradient
is estimated by analysis or other appropriate means. The critical
gradient (Ic) is usually expressed as the ratio of the submerged
unit weight of the soil to the unit weight of water, which results
in the equation (25); $I_c = (G_s-1)/(1+e)$; where G_s is the specific
gravity of the soil particles, and e is the void ratio of the soil
in place.

Laboratory investigations of critical gradient were conducted by
the Bureau for the sandy surficial soils at Calamus Dam. Remolded
specimens of the silty sand were compacted to near field density
and tested using progressively larger upward gradients. Some of
the results were as follows:

a. Measurable dilation and the formation of tiny pipes and surface boils began at a hydraulic gradient of approximately 0.50.

b. Recycling of the upward gradient resulted in the reac-
tivation of the tiny pipes at hydraulic gradients as low as
0.23.

c. Fluidization of this soil could be described as a gradual
process that started at a hydraulic gradient of 0.5 and was
complete near 1.0.

d. The critical gradient (I_c) range computed for the samples
tested was 0.98 to 1.00, which corresponds to a state of
complete failure.

A foundation critical gradient of 0.5 was recommended for use in
the final design of Calamus Dam (29).

The heterogeneity of soils in nature; the progressive and hidden
deterioration that can occur from reservoir fluctuation; and the
sparse data base from appropriate testing justify the continued use
of a conservative Safety Factor (SF = 4) for design of exit gradient protective features.

2. Inadequate protection from high internal gradients will cause
soil particles to migrate from the core and upstream blanket of an
embankment dam, into voids in the foundation and internal drainage

zones. Because the water barrier zones in most embankment dams are
composed of cohesive materials the deterioration will be
progressive, sometimes resulting in lowered density or the formation of pipes. Some examples of situations where unusually high
internal gradients will occur are:

  a. In the upstream blanket or core of a dam underlain by a free
  draining foundation such as coarse alluvium or fractured rock.

  b. In the core of a zoned embankment containing a sloping thin
  core which is undercut by a drainage zone.

3. Excessive pore pressure can contribute to failure modes in the
embankment and the foundation.

  a. Embankment slope instability is a common concern in embankment dams, to which pore pressure effects can be a contributing
  factor. Slope stability is investigated routinely by the Bureau
  (10) on all projects.

  b. Foundation pore pressure can create unstable uplift forces
  on hydraulic structures associated with an embankment dam.

  c. Foundation pore pressure can exert significant uplift force
  on a confining layer of soil immediately downstream of a dam.
  This occurs when there is a lower permeable formation capable of
  transmitting a large percentage of the reservoir head to the
  downstream side of the dam. Failure begins to occur when the
  pore pressure on the bottom of the confining layer exceeds the
  overburden pressure created by the weight of the confining
  layer. The resulting uplift eventually breaches the confining
  layer producing an instantaneous exit gradient in the lower
  transmitting layer. Piping or fluidization failure (as
  described in item 1) can then occur in the lower layer.

The Safety Factor (SF) for this upheaval mode can be expressed
as the ratio of the overburden pressure to the uplift pressure,
which reduces to $SF = G_S t/(1+e)h$; where G_S is the specific gravity of the solids, t the thickness and e the void ratio of the
confining layer soil, and h is the piezometric head at the bottom of the confining layer. Piezometric head is usually determined by analysis or other appropriate means. A safety factor of
2.0 or greater is usually acceptable, depending on the confidence
level of the piezometric head determination.

4. Excess seepage flow is not normally a structural failure mode,
but could be considered in terms of project failure if it results
in inadequate reservoir storage to meet irrigation, domestic water,
or power generation requirements. Downstream flooding or destruc-
tively high ground-water levels could also be considered as project
failure due to excessive seepage.

5. Desicccation of embankment core material without moisture
replenishment leads to moisture content in the core of a dam being
reduced far below the constructed or design intended moisture

content. Desiccation can occur from any combination of: evaporation, plant transpiration and gravity drainage through the dam's built-in drainage features, or through a permeable foundation.

Desiccation is necessarily a long-term process that usually occurs during extended periods of low reservoir level when there is little or no moisture replenishment. The resulting reduction of moisture content in the core material can cause cracking in the core and serious leakage, erosion, and possible failure during subsequent occurances of high reservoir level.

A recent study (13) by the Bureau using a variable saturation finite element model (21) indicated that, for one hypothetical design of a zoned embankment dam on an impermeable foundation, a moisture content reduction of more than 60 percent could occur in the core from end of construction condition to a low reservoir steady-state moisture condition over a period of years. The study further indicated that, even for semiarid climatic conditions, most (approximately 90 percent) of the moisture reduction occurs as the result of drainage via the dam's built-in drainage features. Only a small percentage of moisture loss was attributable to evaporation from the dam's exposed surfaces. The study also shows that moisture content reduction penetrates to a considerable depth in the core.

Criteria for determining the moisture content at which critical desiccation occurs for a given material are not well established. Desiccation cracking is assumed to occur somewhere between the plastic limit and the shrinkage limit for most cohesive soils.

SEEPAGE CONTROL FEATURES

The present design philosophy of the Bureau of Reclamation for seepage control is one of multiple defenses to ensure safety of the embankment structure, and water retention capability of the reservoir. Multiple seepage control features are necessary because of geologic and other factors that may remain unknown from design through the life of the project.

Specific design considerations for seepage control features in embankment dams have been very adequately discussed in other publications (3), (35), and (47). Current Bureau design practice is in close agreement with these guidelines, making it unnecessary that they be discussed in detail in this paper. A brief discussion is given instead.

Embankment zoning normally includes a water barrier zone (core) supported and protected by drainage zones and armor and stability zones. Core materials are selected to be relatively impermeable, nondispersive, and cohesive if possible, with enough plasticity to accommodate some movement without cracking.

The downstream drainage zone (chimney drain) is normally designed with a filter subzone against the core material and highly pervious sand and gravel material for the drain subzone. The filter subzone is

given "critical filter" consideration such as described by Sherard and
others (38), and may have more than one stage for thin core dams. The
chimney drain is extended high enough to accommodate abnormal flows
through cracks that might occur in the upper part of the core.

The upstream stability zone is normally a somewhat pervious back-
fill of material selected to be noncohesive and not prone to cracking,
but containing enough fine grains to plug cracks should they occur in
the core. For thin core dams or dams with low quality core materials,
a specially designed transition subzone is sometimes placed against
the upstream face of the core.

The drainage blanket is a horizontal continuation of the chimney
drain. It must be filter compatible with both the overlying embankment and underlying foundation. It is sized conservatively to accommodate all chimney drain flow and all accumulated water that may be
collected from the foundation and transmit the flow to a low point in
the blanket drain - toe drain system.

Toe drains are designed as a connected part of the embankment
internal drainage system and are sometimes the only drainage feature
needed to mitigate or eliminate seepage exit gradient at the toe of
the dam.

Downstream drainage trenches running parallel to the toe of the dam
are sometimes used when downstream drainage of the foundation is needed
and a low enough discharge point is available. The trench is backfilled
in the drainage zone with pervious soil meeting filter criteria and
incorporates a perforated collector pipe. A drainage trench is being
installed at Foss Dam (43) in Oklahoma to eliminate uncontrolled exit
seepage at the dam and to protect adjoining property from high water
table conditions.

Relief wells are generally used to reduce artesian pressure in confined aquifers beneath a dam or hydraulic structure to tolerable levels
with respect to safety factor against uplift and/or exit gradient
through the confining layer. General design and construction guidance
are provided in references (3) and (17). Relief wells have recently
been installed at two existing dams (Twin Buttes Dam, Texas, and Foss
Dam (43), Oklahoma) to meet dam safety requirements; and are incorporated in the design at a new site presently under construction
(Calamus Dam (31), Nebraska).

Upstream blanketing as a horizontal extension of the embankment
water barrier is intended to lower the seepage gradient in the foundation (and consequently the seepage losses) from reservoir to
tailwater. An upstream blanket is being constructed as one of several
seepage control features at Calamus Dam in Nebraska (30). It is
underlain by silty sands which should provide good filter action.

A geomembrane blanketing the entire reservoir has been used successfully at Mt. Elbert Forebay (46) in Colorado. Special earth

bedding and cover layers were placed with a 1-inch or smaller particle size restriction to prevent perforation of the membrane.

Slurry wall cutoff is one of the most effective methods of creating a water barrier in the foundation of a dam. It is used by the Bureau when extension of the embankment cutoff trench to an impermeable foundation is uneconomical. The technology is well described in reference (49).

A slurry wall is another one of the seepage control features incorporated into the design of Calamus Dam (30). It is fully penetrating where the top of foundation rock is shallow and partially penetrating in the deeper areas.

Diaphragm walls are usually constructed of concrete sometimes using patented processes and equipment of which reference (39) is an example. They are seldom used in embankment dams because of the cost. They are an effective means of resolving a potentially unsafe embankment dam situation involving both dam and foundation, such as at Wolf Creek Dam (23). Diaphragm wall test sections are planned for Fontenelle Dam in Wyoming to eventually correct a situation where the embankment core has been softened by piping into a fractured rock foundation.

Grout curtains are used by the Bureau on many dams. As a seepage cutoff feature, their effectiveness varies greatly depending on geologic conditions. Neat cement grout mixes are the most common grouts used by the Bureau. Chemical grouting is seldom used by the Bureau because of environmental and economic concerns. General Bureau practice in grouting is discussed in reference (7).

Drainage tunnels from which a series of drain holes can be drilled are sometimes necessary to relieve seepage pressure from rock abutments and remove seepage flows from the embankment-foundation contact. Consideration should always be given to possible adverse effects of increasing the gradient through the embankment core when such drainage of the contact occurs. Drainage tunnels are presently being constructed at Soldier Creek Dam (48) in Utah to control potentially adverse seepage at the dam/abutment contact.

Downstream seepage berms are sometimes used by the Bureau to counteract high exit gradient or uplift pressure at the toe of a dam. A permeable downstream berm was constructed at Twin Lakes Dam in Colorado when downstream piping occurred during first filling of the reservoir.

CONCLUSION

Most seepage-related problems in embankment dam design involve deficiencies in the natural foundation, while much better control of seepage in the embankment has been achieved because it is manmade.

One area of continued Bureau (13) interest is the development of better control over desiccation cracking (41) (42) of the embankment core, in terms of the following:

1. Confirmation of the major mechanisms causing desiccation
2. Confirmation of the depth of penetration of desiccation
3. Design considerations to minimize desiccation
4. Design features for adding supplemental moisture
5. Material test procedures to establish critical state for desiccation cracking
6. Instrumentation to monitor desiccation

APPENDIX I - REFERENCES

1. Adhya, K. K., "Seepage Analysis, McPhee Dam and Reservoir," TM No. VII-224-B, Dolores Project, Bureau of Reclamation, Denver, Colo., 1980 (unpublished)
2. Bureau of Reclamation, "Earth Manual," Second Edition, Denver, Colo., 1974
3. Bureau of Reclamation, "Design of Small Dams," Second Edition, Denver, Colo., 1977
4. Bureau of Reclamation, "Groundwater Manual," First Edition, Denver, Colo., 1977
5. Bureau of Reclamation, "Seepage Analysis for the Stability of Narrows Dam," Denver, Colo., Memorandum dated November 8, 1977 (unpublished)
6. Bureau of Reclamation, "Drainage Manual," First Edition, Denver, Colo., 1978
7. Bureau of Reclamation "Policy Statements for Grouting," ACER Technical Memorandum No. 5, Denver, Colo., 1984
8. Casagrande, A., "Seepage Through Dams," Contributions to Soil Mechanics 1925-40, Boston Society of Civil Engineers, Boston, Mass., 1940
9. Cedegren, H. R., "Seepage, Drainage and Flow Nets," John Wiley Sons, Second Edition, 1977
10. Chugh, A. K., "Slope Stability Analysis Program SSTAB2," User Information Manual, Bureau of Reclamation, Denver, Colo., 1981
11. Chugh, A. K., and H. T. Falvey, "Seepage Analysis in a Zoned Anisotropic Medium by the Boundary Element Method," International Journal for Numerical and Analytical Methods in Geomechanics, Vol. 8, 1984
12. Cobb, J. E., "Seepage Analysis Report, Stage 3 Specifications," TM No. V-222-B-1-g, Calamus Dam Design Summary, Bureau of Reclamation, Denver, Colo., 1983 (unpublished)
13. Cobb, J. E., and C. L. Mantei, "Variable Saturation Studies," TM No. BR-230-41, Brantley Dam Design Summary, Bureau of Reclamation, Denver, Colo., 1984 (unpublished)
14. Committee on the Safety of Existing Dams "Safety of Existing Dams: Evaluation and Improvement," National Research Council, National Academy Press, Washington, D.C., 1983
15. Corps of Engineers, "Soil Mechanics Design, Seepage Control," EM 1110-2-1901, U.S. Army, Office of the Chief of Engineers, Washington, D.C., 1952
16. Corps of Engineers, "Investigation of Underseepage and its Control, Lower Mississippi River Levees," TM No. 3-424, U.S. Army, Waterways Experiment Station, Vicksburg, Miss., 1956
17. Corps of Engineers, "Design of Finite Relief Well Systems," EM 1110-2-1905, U.S. Army, Office of the Chief of Engineers, Washington, D.C., 1963

18. Coprs of Engineers, "Permeability of Foundation Soils," Report
    No. 1, Arkansas River Lock and Dam Stuctural Studies, U.S. Army
    Engineer District, Little Rock, Ark., 1970
19. Corps of Engineers, "Design and Construction of Levees,"
    EM 1110-2-1913, U.S. Army Office of the Chief of Engineers,
    Washington, D.C., 1978
20. Corey, A. T., "Mechanics of Heterogeneous Fluids in Porous Media,"
    Water Resource Publications, Fort Collins, Colo. 1977
21. Davis, L. A., and S. P. Neuman, "Documentation and Users Guide:
    UNSAT2-Variably Saturated Flow Model," NUREG/CR-3390,
    WWL/TM-1791-1, Water Waste and Land, Inc., Fort Collins, Colo.,
    1983
22. Davis, R. D., "Seepage Model Parameter Study of Foundation Design
    Alternatives," TM No. BA-222-26, Batu Dam Design Summary, Bureau
    of Reclamation, Denver, Colo, 1982 (unpublished)
23. Fetzer, C. A., "Wolf Creek Dam, Remedial Work, Engineering
    Concepts, Actions, and Results," Transactions of the Thirteenth
    International Congress on Large Dams, Vol. 2, New Delhi, India,
    1979
24. Freeze, R. A., "Influence of the Unsaturated Flow Domain on
    Seepage Through Earth Dams," Water Resources Research, Vol. 7,
    No. 4, 1971
25. Harr, M. E., "Groundwater and Seepage," McGraw-Hill Book Co., 1962
26. Hirschfeld, S. D., and S. J. Poulos, ed., "Embankment Dam
    Engineering," Casagrande Volume, John Wiley and Sons, 1973
27. Johnson, A. I., "Specific Yield - Compilation of Specific Yields
    for Various Materials, "Water Supply Paper 1662-D, U.S.
    Geological Survey, 1967
28. Justin, J. D., J. Hinds, and W. P. Creager, "Engineering for
    Dams," Vol. III, John Wiley and Sons, 1945
29. Long, J. C. S., J. S. Remer, C. R. Wilson, and P. A. Witherspoon,
    "Porous Media Equivalents for Networks of Discontinuous
    Fractures," Water Resources Research, Vol. 18, No. 3, 1982
30. McAlexander, E. L., T. N. McDaniel and C. L. Mantei, "Calamus Dam:
    Design and Analyses," Transactions, Fourteenth International
    Congress on Large Dams, Vol. II, Rio de Janeiro, Brazil, 1982
31. McAlexander, E.L., and W. O. Engemoen, "Design and Monitoring for
    Seepage at Calamus Dam" ASCE, (this session)
32. Milligan, V., "Field Measurement of Permeability in Soil and
    Rock," ASCE, Proceedings of the Conference on In Situ
    Measurement of Soil Properties, North Carolina State University,
    Vol. II, 1975
33. Morris, D. A., and A. I. Johnson "Summary of Hydrologic and
    Physical Properties of Rock and Soil Materials, as Analyzed by
    the Hydrologic Laboratory of the U.S. Geological Survey
    1948-60," Water Supply Paper 1839-D, U.S. Geological Survey,
    1967
34. Neuman, S. P., R. A. Feddes, and E. Bresler, "Finite Element
    Simulation of Flow in Saturated-Unsaturated Soils Considering
    Water Update by Plants," Third Annual Report, Project
    No. ALO-SWC-77, Technion Haifa, Israel, 1974
35. Papagianakis, A. T., and D. G. Fredlund, "A Steady State Model for
    Flow in Saturated-Unsaturated Soils," Canadien Geotechnical
    Journal, Vol. 27, No. 3, 1974
36. Powers, J. P, "Construction Dewatering" John Wiley and Sons, 1981

37. Sherard, J. L., R. J. Woodword, S. F. Gizienski, and
    W. A. Clevenger, "Earth and Earth-Rock Dams," John Wiley and
    Sons, 1962
38. Sherard, J. L., L. P. Dunnigan, and J. R. Talbot, "Filters for
    Silts and Clays," ASCE, Journal of Geotechnical Engineering,
    Vol. 110, No. 6, 1984
39. SOLETANCHE, "Works and References," Nanterre Cedex, France
40. Stevens, D. B., S. P. Neuman and others, "In Situ Determination
    of Hydraulic Conductivity in the Vadose Zone Using Bore Hole
    Infiltration Tests," Technical Completion Report, Project
    No. 1423648, New Mexico Water Resources Research Institute,
    New Mexico State University, Los Cruces, New Mex. 1983
41. Stevenson, J. C., R. Smith, and C. E. Stearns, "Cracking of Earth
    Dams in Arizona," Report of Crack Study Team, U.S. Soil
    Conservation Service, Portland, Ore., 1978 (unpublished)
42. Talbot, J. R., "Cracking of Dams in Arid Regions," ASAE, 1980
    Winter Meeting, Paper No. 80-2548, 1980
43. Torres, R. L., "Seepage Analysis for Modification of Foss Dam"
    TM No. FO-VI-222-1, Foss Dam Modification Design Summary, Bureau
    of Reclamation, Denver, Colo., 1982 (unpublished)
44. Tracy, F. T., "A Three-Dimensional Finite Element Program for
    Steady State and Transient Seepage Problems," Misc. Paper
    K-73-3, U.S. Army Engineer Waterways Experiment Station,
    Vicksburg, Miss., 1973
45. Tracy, F. T., "A Plane and Axisymmetric Finite Element Program
    for Steady State and Transient Seepage Problems," Misc. Paper
    K-73-4, U.S. Army Engineer Waterways Experiment Station,
    Vicksburg, Miss., 1973
46. USCOLD "Mt. Elbert Forebay Reservoir," Newsletter, March 1981
47. Wilson, S. D., and R. J. Marsal, "Current Trends in Design and
    Construction of Embankment Dams," ASCE, New York, 1979
48. Wiltshire, R. L., "Soldier Creek Dam Foundation Drainage," ASCE,
    (this session)
49. Xanthakos, P. P., "Slurry Walls," McGraw-Hill, 1979
50. Yang, S. S., "Preliminary Dewatering Studies of Calamus Damsite,"
    TM No. V-222-B-1-I, Calamus Dam Design Summary, Bureau of
    Reclamation, Denver, Colo., 1982
51. Zangar, C. N., "Theory and Problems of Water Percolation,"
    Engineering Monographs No. 8, Bureau of Reclamation, Denver,
    Colo., 1953

SUBJECT INDEX
Page number refers to first page of paper.

AUTHOR INDEX

Page number refers to first page of paper.